GAME CHANGERS

www.energygamechangers.org

A production of the Shultz-Stephenson Task Force on Energy Policy at Stanford University's Hoover Institution and the Massachusetts Institute of Technology Energy Initiative.

HOOVER INSTITUTION

SHULTZ-STEPHENSON TASK FORCE ON
Energy Policy

MIT Energy Initiative

Jeremy Carl
Research Fellow,
The Hoover Institution
Director of Research,
Shultz-Stephenson Task Force
on Energy Policy

David Fedor
Research Analyst,
Shultz-Stephenson Task Force
on Energy Policy,
The Hoover Institution

David Slayton
Research Fellow,
Shultz-Stephenson Task Force
on Energy Policy,
The Hoover Institution

George P. Shultz
Chairman of the Shultz-Stephenson
Task Force on Energy Policy and
Thomas W. and Susan B. Ford
Distinguished Fellow at Stanford
University's Hoover Institution

Robert C. Armstrong
Director of the MIT Energy Initiative
and Chevron Professor of chemical
engineering

Louis Carranza
Associate Director,
MIT Energy Initiative

Rebecca Marshall-Howarth
Director of Editorial Services
and Publications,
MIT Energy Initiative

Francis O'Sullivan
Director of Research and Analysis,
MIT Energy Initiative
Lecturer, MIT Sloan School
of Management

GAME
CHANGERS
energy on the move

Five R&D efforts from American universities
that are offering a cheaper, cleaner, and
more secure national energy system

Edited by George P. Shultz
and Robert C. Armstrong

HOOVER INSTITUTION PRESS
STANFORD UNIVERSITY STANFORD, CALIFORNIA

The Hoover Institution on War, Revolution and Peace, founded at Stanford University in 1919 by Herbert Hoover, who went on to become the thirty-first president of the United States, is an interdisciplinary research center for advanced study on domestic and international affairs. The views expressed in its publications are entirely those of the authors and do not necessarily reflect the views of the staff, officers, or Board of Overseers of the Hoover Institution.

www.hoover.org

Hoover Institution Press Publication No. 656

Hoover Institution at Leland Stanford Junior University, Stanford, California 94305-6010

About the cover: Microsystem-enabled photovoltaics—or "solar glitter"—developed at Sandia National Laboratories reduce crystalline silicon material needs by a factor of ten without sacrificing solar conversion efficiency. It joins other fruitful research efforts at universities and in industry from around the country in yielding new technological options for cheaper, more secure, and cleaner power generation.

For more on energy R&D from the lab, visit *www.energygamechangers.org.*

First printing 2014
20 19 18 17 16 15 14   7 6 5 4 3 2 1

Manufactured in the United States of America

The paper used in this publication meets the minimum Requirements of the American National Standard for Information Sciences—Permanence of Paper for Printed Library Materials, ANSI/NISO Z39.48-1992. ∞

Cataloging-in-Publication Data is available from the Library of Congress.
ISBN: 978-0-8179-1825-5 (pbk. : alk. paper)
ISBN: 978-0-8179-1826-2 (epub)
ISBN: 978-0-8179-1827-9 (mobi)
ISBN: 978-0-8179-1828-6 (PDF)

CONTENTS

AVAILABLE TODAY IT-enabled "smart" oil fields now make mountains of computational data available in real time to project engineers in the field or at distant control centers.

NEAR AT HAND University of Texas's downhole electromagnetic monitoring of frac proppants in a shale formation can greatly enhance fracture efficiency at low cost.

NEAR AT HAND A research team at the University of Texas has developed a novel hydrophobic membrane that resists clogging and enables fresh water savings of 50 percent during the fracing process.

NEAR AT HAND Stanford's ambient seismic oilfield monitoring technology could help oil companies improve recovery techniques and let drillers more effectively monitor existing oil fields.

ON THE HORIZON MIT's millimeter-wave rock-vaporizing directed energy beams could replace drill bits to access underground energy resources.

ON THE HORIZON University of Texas researchers have shown that hot brines saturated with dissolved methane in the Gulf of Mexico could be used to recover vast amounts of methane and geothermal energy, and also be a store of injected carbon dioxide.

AVAILABLE TODAY High-efficiency monocrystalline rear-junction silicon cells have over three decades grown from Stanford graduate student research into a multibillion-dollar business with dramatically reduced manufacturing costs.

NEAR AT HAND Stanford's nanocrystalline-silicon shells improve thin solar panel light absorption, reducing materials usage and processing costs.

NEAR AT HAND MIT's three-dimensional solar cell arrangements could reduce installation cost and increase power output per base area.

ON THE HORIZON MIT's thin-film organic polymer flexible solar cells, created with moderate temperatures and no liquids, have been printed on tissue paper, textiles, and even plastic food wrap.

ON THE HORIZON University of Texas's printable inorganic thin-film flexible solar cells have reached efficiencies of nearly 4 percent without the need for expensive high-temperature manufacturing processes.

ON THE HORIZON A University of Michigan team is analyzing the potential of photovoltaic technology based on semitransparent organic material that could be incorporated into conventional built environment surfaces.

AVAILABLE TODAY A new generation of research into compressed air energy storage technology reduces costs and improves efficiency by changing how these systems handle heat during compression and expansion.

AVAILABLE TODAY USC and Caltech's direct methanol fuel cell technology has proven commercially useful for long-lasting, low-power electricity delivery in a wide variety of distributed power applications.

NEAR AT HAND MIT researchers have unraveled the properties of a "superlattice" material structure that improves the reactivity of fuel cell electrodes.

NEAR AT HAND Research at Stanford on multiwalled carbon nanotubes and graphene has found a way to reduce the cost of fuel cells by replacing platinum catalysts.

ON THE HORIZON MIT's liquid-metal batteries employ a novel "reverse smelting" process to lower the costs and increase the longevity of large-scale energy storage.

ON THE HORIZON Stanford's low-cost crystalline copper hexacyanoferrate large-scale battery electrode lasts for 40,000 cycles of charging and discharging.

ON THE HORIZON A Stanford research team's simplified lithium polysulfide membrane-free flow battery performs well over more than 2,000 discharge cycles.

AVAILABLE TODAY MIT's carbon nanotube enhanced ultracapacitor, now commercially available, stores twice as much energy as conventional alternatives and delivers seven to fifteen times more power.

AVAILABLE TODAY The origins of lithium-ion batteries, now ubiquitous in mobile electronics and increasingly so in electric vehicles, can be traced to university and industry research in the 1970s and 1980s.

NEAR AT HAND Stanford's novel electrodes made of silicon nanoparticles and conducting polymer hydrogel dramatically improve the performance of lithium-ion batteries.

LEDs convert electricity to ultraviolet light by up to
four times.

ON THE HORIZON An international research team has
demonstrated a revolutionary electrically driven polariton
laser that could significantly improve the efficiency of
lasers.

ON THE HORIZON An interdisciplinary research team at
Stanford has created a device that tames the flow of
individual photons with synthetic magnetism.

ABBREVIATIONS

ABE	Acetone, Butanol, and Ethanol
CdTe	Cadmium telluride
CDEW	Counter-Directed Energy Weapon
CFL	Compact Fluorescent Lamp
CIGS	Copper Indium Gallium Diselenide
CO_2	Carbon dioxide
CUBE	Consolidated Utility Base Energy
CVD	Chemical Vapor Deposition
DARPA	Defense Advanced Research Projects Agency
DETER	Defense Technology Experimental Research
DoD	Department of Defense
DOE	Department of Energy
EM	Electromagnetic
EOR	Enhanced Oil Recovery
EPRI	Electric Power Research Institute
ExFOB	Experimental Forward Operating Base
FOB	Forward Operating Base
GVL	Gamma-Valerolactone
HCCI	Homogeneous Charge Compression Ignition

IED   Improvised Explosive Device
IT   Information Technology
HEL   High-Energy Laser
JPL   Jet Propulsion Laboratory
LED   Light-Emitting Diode
LNG   Liquefied Natural Gas
MITEI   MIT Energy Initiative
MPPT   Maximum Power Point Tracking
MTC   Mobility Transformation Center
MTVR   Medium Tactical Vehicle Replacement
oCVD   oxidative Chemical Vapor Desposition
OECD   Organization for Economic Cooperation and
           Development
OLED   Organic Light-Emitting Diode
PETE   Photon-Enhanced Thermionic Emission
PV   Photovoltaic
SLAC   Stanford Linear Accelerator Center
SVP   Smart Voyage Planning
SWIPES   Soldier Wearable Integrated Power System
TEPCO   Tokyo Electric Power Company

PREFACE

The idea behind Game Changers is that there are energy
technologies that are well within reach.

—Susan Hockfield, *MIT president emerita*

Revolution is a strong word that is genuinely applicable to the
field of energy today, both in the way it is produced and the way
it is used. The challenges ahead are compelling and the sooner
we take advantage of our resources and our capacity for innovation, the brighter our future will be.

All nations, particularly those with advanced economies like
the United States, need reliable and inexpensive energy to propel economic growth. We also need to produce and use energy
in a way that is compatible with our national security interests.
It is becoming increasingly obvious that we need to produce more
energy where we use it because interruptions from natural causes
or hostile cyber attacks are likely to occur more frequently.

Then there is the environment—the air we breathe and the
climate we influence. We are impressed with the science providing careful analysis about the warming of our climate. The new
ocean that is being created in the Arctic for the first time since the
Ice Age with important consequences—positive and negative—is
observable evidence that the planet is growing warmer.

What can be done? As Energy Secretary Ernie Moniz said
when he was director of the MIT Energy Initiative (MITEI),

there is tremendous research capacity in the United States and elsewhere to carry out aggressive work on this important agenda. The United States needs to exploit this capacity by devoting scientific, engineering, and entrepreneurial resources to the energy area on an unprecedented scale.

The fruits of this research and development, combined with the creative juices of the American entrepreneurial culture, will shape our future. Universities can lead the way in this research and businesses can commercialize and scale the good, underlying ideas, which are the key.

Recognizing this reality, we at Stanford and MIT have joined together to recognize new, important, and potentially game-changing energy technologies. Twenty-five scientists from our two universities convened at Stanford's Hoover Institution in 2011 to review specific projects. The group met again the following year at MITEI. Some of the most compelling results from those meetings were presented at a conference cosponsored by Hoover and MITEI in Washington, DC, in March 2013.

The purpose of this book is to highlight those and other genuine energy game changers—new ideas for how to produce and use energy in a careful way. In addition to spotlighting innovations from our two universities, we include contributions from other leading research centers at UT Austin, UC Berkeley, Caltech, Georgia Tech, the University of Michigan, and the University of Southern California.

We have organized these ideas into three categories: technologies that have been commercialized, technologies on the cusp of becoming operational, and potential blockbuster technologies down the road that hold real promise of major importance. All of these ideas are portrayed against the backdrop of our three key objectives: national security, economic well-being, and an improved environment.

We start by taking note of a development that is rapidly changing the availability of natural gas and crude oil—a technology known as hydraulic fracturing. This technology has already had a huge impact on the availability of natural gas in the United States and promises important developments on a global scale. The scale that the natural gas resource has reached in just a few years underlines another reality: although the energy market is massive, a scalable and competitive development can make an impact in a reasonably short period of time. This dramatic progress shows how research and development can interact with entrepreneurial capability, and then larger corporate America, to produce stunning results.

We believe that we are on the brink of a golden age in the field of energy. There is a far greater mass of first-class science and engineering being applied to energy now than ever before in the United States and elsewhere. Game-changing energy innovations are the engine toward a bright future. But while the possibilities for dramatic new game changers are endless, we are not exploiting those opportunities. We in the United States have been on an unfortunate roller coaster for the last fifty years: a crisis motivates research efforts, which then recede when that crisis ends. We need to rev up the engine of energy innovation in our country, and this time, we cannot allow that engine to stall.

—George P. Shultz
Chairman of the Shultz-Stephenson Task Force on Energy Policy
and Thomas W. and Susan B. Ford distinguished fellow at
Stanford University's Hoover Institution

—Robert C. Armstrong
Director of the MIT Energy Initiative and
Chevron professor of chemical engineering

ACKNOWLEDGMENTS

The Game Changers initiative grew out of conversations on the emergence of new energy research in the early 2000s that we developed with Susan Hockfield while she was president of MIT. Words came to fruition alongside the robust development of the Global Climate and Energy Project, MITEI, and Stanford's Precourt Institute for Energy. We would like to thank directors Sally Benson, Ernie Moniz, and Lynn Orr for their close involvement in helping to assemble a host of affiliated faculty researchers over three years and three joint conferences. MIT President Rafael Reif and Stanford President John Hennessey also made great personal contributions over the years through their participation in the Game Changers effort.

We also thank the faculty and staff at universities around the country who helped to identify and report on the numerous energy research and development efforts featured here. They volunteered many more compelling results than we had room to include:

- Thomas Edgar, Michael Webber, and Carey King at the University of Texas at Austin Energy Institute;

- Mark Barteau and Amy Mast at the University of Michigan Energy Institute;
- Paul Wright at the University of California at Berkeley Energy and Climate Institute, along with Chris Somerville at the Energy Biosciences Institute;
- Tim Lieuwen and Allison Davis at Georgia Tech's Strategic Energy Institute with John Toon at Georgia Tech Research News;
- Nate Lewis at the Caltech Joint Center for Artificial Photosynthesis with Dave Zobel at Caltech Media;
- Donald Paul at the University of Southern California's Energy Institute; and
- Emily Carter at Princeton's Andlinger Center for Energy and the Environment.

Closer to home, we are indebted to those at Stanford and MIT whose reporting on campus energy innovation appears here. At MIT, Nancy Stauffer at MITEI and David Chandler and Anne Trafton at the MIT News Office. At Stanford, Mark Shwartz of the Stanford Precourt Institute for Energy, Louis Bergeron and Thomas Sumner at the Stanford News Service, Andrew Myers at the School of Engineering, Mike Ross, Andy Freeberg, and Glennda Chui of the SLAC National Accelerator Laboratory, and Julia Barrero for the Stanford Peninsula Press.

Jeremy Carl, David Slayton, and David Fedor at Stanford University's Hoover Institution collaborated with Melanie Kenderdine, Francis O'Sullivan, Rebecca Marshall-Howarth, Debi Kedian, and Louis Carranza at the MIT Energy Initiative to organize the Game Changers events and help produce this publication. Meanwhile, Ambassador Thomas Stephenson has supported the work of Hoover Institution's Shultz-Stephenson Task Force on Energy Policy on Game Changers with both his generosity and his counsel.

Finally, we would like to specially acknowledge those who have sponsored energy research and development at universities around the country. At Stanford and MIT in particular, we thank:

- The US federal government, including the Department of Energy (DOE), the Department of Defense (DoD), their essential ARPA-E and Defense Advanced Research Projects Agency (DARPA) programs, and the Congress members, citizens, and taxpayers who make their work possible;
- Industry sponsors and collaborators including, at Stanford, the Global Climate and Energy Project's ExxonMobil, General Electric, Schlumberger, Toyota, DuPont, and Bank of America alongside many other generous industry affiliates and, at MIT, founding members of MITEI, including BP, Eni, Saudi Aramco, Shell, and the other benefactor organizations that provide sustaining support; and
- Selfless individuals and civil foundations such as Jay Precourt, Tom Steyer and Kat Taylor, and the King Abdullah University of Science and Technology whose early support helps researchers pursue new ideas that might otherwise be left on the bench.

Together with the faculty and student researchers in the lab, their support underpins the energy innovations of today and those of tomorrow.

—George P. Shultz
*Chairman of the Shultz-Stephenson Task Force on Energy Policy
and Thomas W. and Susan B. Ford distinguished fellow at
Stanford University's Hoover Institution*

—Robert C. Armstrong
*Director of the MIT Energy Initiative and
Chevron professor of chemical engineering*

CHAPTER 1

NATURAL GAS
FROM SHALES

Recent advancements in horizontal drilling and hydraulic fracturing of shale formations have changed the landscape of US
oil and gas production. There were over 48,000 oil and gas
wells drilled in the United States alone in 2012. Fifty percent
of these new wells were horizontal wells; most of them drilled
in unconventional oil and gas plays; and over 95 percent of
them hydraulically fractured. The result of all this is now well
known: dramatically increased domestic production of natural gas. These subsurface innovations—enabled in part by new
technologies such as downhole imagery, microseismic imaging,
and slick water fracturing—have both driven down natural gas
prices and strengthened the contributions of the gas sector itself
to the US economy.

Lower natural gas prices have led to significant environmental improvements. For the first time in decades, the
US electric grid in mid-2012 was supplied by approximately
equal shares of gas- and coal-fired power generation, though
coal's share recently increased in part due to rising gas prices.
Gas-for-coal substitution of course is attractive from a climate

change perspective (and, in fact, according to several recent studies, coal-to-gas switching dwarfs the marginal contributions of renewable energy), but it also has huge—and largely overlooked—local environmental and health advantages.

For example, the respiratory impacts of local pollution from coal-fired power generation are routinely estimated to result in approximately 10,000 statistical lives lost in the United States each year. At the same time, low natural gas prices enabled by the uptake of shale gas technologies have helped drive nearly a 20 percent decline in coal-fired generation in the United States. At even a conservative estimate of attribution, new shale natural gas production is helping to save 1,000 lives per year, ongoing each year, across the United States. Further research and monitoring on potential adverse environmental impacts of natural gas extraction are advisable (as it would be for any source of energy), but these costs should be weighed against the real economic environmental and health benefits that the country is already enjoying from expanded gas production today.

Apart from gas, it is also worth noting that horizontal drilling technologies, along with ultradeepwater offshore drilling, have also helped increase US domestic oil production in the lower 48 states after decades of gradual declines. Along with improved efficiency of oil use elsewhere in the economy, the Energy Information Administration estimates that this has helped reduce US net oil imports to 42 percent, down from a peak of over 60 percent in 2006. Also, with advancements in Canadian oil sands production, only about 22 percent of the United States' 2011 oil needs were met from suppliers outside the Western Hemisphere. This is helping to improve US energy security against the risk of severe global supply disruptions.

When you have an advanced technology available, the policy decisions become much easier.

—Julio Friedmann, *Lawrence Livermore National Laboratory chief energy technologist**

Claiming credit for delivering the US shale gas boom has become an energy policy parlor game, and the truth is that today's shale gas boom has many fathers. Mitchell Energy's tenacious trial-and-error experimentation with new and unproven field techniques persevered despite years of subpar returns. Serendipity—high gas prices, well-timed supply contracts, and convenient geologies—allowed this experimentation to continue long enough for costs to be gradually driven down. The United States' private-property rights regime, almost unique in the world in terms of its aggressive assignment of private property rights over mineral resources, offered an exit opportunity to compensate for early investment risks in a sector in which operational advances often spill over. A synergistic corporate acquisition combined key know-how—Mitchell's shale slick water fracturing techniques with Devon Energy's gas horizontal drilling capabilities—that made the process economic throughout gas basins outside of Texas's Barnett. And existing gas gathering and pipeline infrastructures, built over decades, ultimately provided ready legs to achieve today's shale gas-production scale.

Crucially, many of these contributions were enabled by years of industry, consumer, and government-sponsored research and collaboration. Some scientific and technical contributions were dead ends, and others took decades before their value was fully recognized. But without this R&D, permits for siting liquefied natural gas (LNG) import terminals—and not today's backlog of applications for hotly contested export licenses—would likely still be on top of the US energy policy agenda. Two recent investigations from the Breakthrough Institute and Resources for the Future trace this story of shale gas technology's development:

The story begins with research conducted by national labs, universities, and the private sector under DOE's Eastern Gas Shales Project, which ran from 1978 to 1992 in the Devonian shales of the Appalachians and ultimately aimed to increase production by improving the technology available for this geology. The program applied horizontal drilling, which had become commercial in oil fields through the 1980s, to Devonian gas shales. Massive hydraulic fracturing was similarly applied to gas shales using the techniques already proven in tight gas formations. The program also pioneered the use of foam fracturing, which Mitchell Energy would later use—and ultimately discard in favor of simpler water fracturing—in its attempts to "crack the Barnett." Other related public- and private-sponsored research contributions included:

- The early development of more effective diamond-studded drill bits through a public-private federal government collaboration with GE.
- The adaptation of downhole passive microseismic fracture monitoring to gas shales—a technology that originated in work done by Los Alamos National Lab for geothermal energy in the 1970s—through a collaboration between DOE and the industry-supported Gas Research Institute.
- Industry-led advances in 3D seismic imaging that were enabled by high-performance computing. This high-resolution subsurface mapping was particularly important

to identify and recover from previously uneconomic distributed shale deposits.

Mitchell Energy was, however, the most important singular force in applying new technologies and operations that would commercialize shale gas production in the United States. Mitchell plowed a quarter of a billion dollars of in-house R&D budgets over two decades into bringing down the cost of shale gas production in north Texas's Barnett formation. Starting with a "gelled" water frac process previously developed through a DOE collaboration in east Texas, Mitchell relentlessly iterated to gradually bring down costs per well by simplifying the process. First, the nitrogen assist was removed from the original process, followed by the substitution of a cheaper low-quality sand proppant, removal of pre-frac acid treatments, and finally even the gel itself—all without significantly reducing effectiveness. This learning-by-doing process culminated in the deployment of the "slick water fracturing" technology (originally developed by Union Pacific Railroad for tight gas formations and, before it, Exxon for oil wells) for shales, reducing well stimulation costs by half and overall well costs by hundreds of thousands of dollars.

Mitchell Energy was rewarded for its early risk taking when the substantial shale acreages it had built up throughout the Barnett during the long applied R&D process (informed by public- and private-sector geologic characterizations) were recognized for their newfound commercial value. And following the multibillion-dollar Mitchell-Devon Energy merger in 2002, shale gas development took off. The seemingly disparate technologies that had been explored over the past twenty to thirty years of R&D fell into place as enablers of commercial operations.

Mitchell's slick water fracturing was combined with Devon's expertise in horizontal drilling, something that Mitchell had been unable to accomplish on its own; microseismic fracture mapping was successfully used to monitor stimulated shale gas well performance; and 3D seismic was increasingly used to identify new shale gas resources and guide low-tolerance horizontal drilling operations. Finally, a period of sustained high domestic gas prices provided headroom to continued development, eventually leading to the widespread use we see today across the United States. This stands today as one of the strongest examples of a public-private research partnership from many disparate strands, enabling the growth and development of a critical energy technology.

AVAILABLE TODAY   IT-enabled "smart" oil fields

Data-hungry oil and gas companies have long been early adopters of IT. The first digital signal-filtering methods developed at MIT in the 1950s were actually used to improve underwater seismic surveys. Adoption of computing technology really took off in the 1960s with the transition from vacuum tubes to transistors as the industry migrated to digital systems. Texas Instruments actually grew out of a geophysical services consultancy, oil majors regularly commissioned custom IBM machines for decades, and today BP operates the largest commercial supercomputer.

The past decade, however, has seen a sea change not just in computing hardware, but also in the software and supporting operational capability available to oil producers. So-called "integrated asset management" systems—or more simply, smart oil fields—now make the mountains of computational data (drilling logs, seismic maps, well flow models, production levels, and equipment status) available in real time to project engineers in the field or distant control centers. The result? Increased recovery rates, improved safety, less downtime, and more efficient capital deployment. Chevron, for example, uses integrated IT today to standardize reservoir operations across its hundreds of fields globally that use waterflooding-based recovery.

The multidisciplinary advances to enable smart oilfield systems have drawn from both industry and academia. At the University of Southern California, for example, researchers in computer science and electrical engineering have described novel databases for better on-site access to existing field scenario analyses. Such systems—developed in partnership with industry—even allow field engineers to update existing predicted models in real time based on actual well production. Elsewhere, USC researchers in informatics, with original funding from DARPA and the US Air Force, have developed their research into overlaying semantic and text-based information onto geospatial maps—a system then called TILES—into commercially available geographic information system software that is used across industries.

In the five years since the start of the hydraulic fracturing boom, the United States has not only become the global leader in the production of natural gas from shales, but also the fastest-growing producer of oil. This means new wells in new landscapes—many of them populated, and many of them previously untouched by upstream oil and gas operations. Ongoing research seeks to improve understanding and mitigation of the industry's environmental effects—both in general and those that are particular to fracturing itself—including atmospheric, surface, and subsurface water contamination.

We don't actually know if we're doing a good job right now. We're maybe recovering between eight and thirteen percent of the resource. We don't know if the technologies we're using today are impairing future recovery of future resource. We don't know whether or not we're actually shooting ourselves in the foot today in terms of a national asset which has not been measured well and has not been

developed in a way that's thinking about the long-term potential value.
A lot of the potential allure of recovery—getting the flow paths right,
making sure that we don't damage aquifers, and a number of other
things—boils down to the ability to make and control fractures
in the subsurface and the coupled hydrology that comes with
them. That geo-mechanical problem is one that has been grossly
underserved in a federal research program for a long, long time.

—Julio Friedmann, *Lawrence Livermore National Laboratory chief
energy technologist**

Better understanding of subsurface-induced fracturing
dynamics could also help improve hydrocarbon production efficiency and resource management. Moreover, unconventional
domestic oil and gas asset estimates have not kept pace with new
production growth. Work is currently ongoing to develop more
credible geologic characterizations of these new resources and
reserves, with academic researchers playing a significant role.

The need for fundamental and chemical mechanisms controlling flow
from the nanoscale to the basin scale....This kind of research is going
on in a number of universities around the country and the question of
how much gas we're actually going to be able to recover boils down
to nanoscale processes that are not yet well constrained. We have
these large maps. We draw circles around large basins, but actually
on an individual well basis, we don't understand these processes
well enough to answer the question of whether or not these wells
are going to last five years or they're going to last twenty-five years.
And this research is absolutely essential. . . . the private sector
is very heavily involved in all these things and it's a combined
responsibility to move the ball forward.

—Mark Zoback, *Stanford University Benjamin M. Page professor
of geophysics*

The potential to increase recovery from existing fields—so no new surface infrastructure—is about 5–15 percent enhanced recovery. Subsurface micro and nanosensors—a third remote sensing platform, different from well logging and seismic—are being developed to address this EOR [enhanced oil recovery] target.

—Scott Tinker, *Bureau of Economic Geology director at the University of Texas at Austin*

NEAR AT HAND   **University of Texas's downhole electromagnetic monitoring of proppants**

A UT Austin researcher has shown that strategically timing successive fracturing operations in a shale formation can greatly enhance fracture efficiency.

Hydraulic fracturing in shale formations induces a large number of microseismic events in the rock surrounding the production zone. These induced fractures are so narrow that they generally do not contain proppant from the main hydraulic fracture and therefore do not produce gas. Induced, unpropped fractures do, however, result in a "stress shadow" that affects the direction and extent of propagation of subsequent fractures. One unwanted consequence of this is the propagation of subsequent fractures into the unpropped fractures induced earlier, leading to waste of frac fluid and proppant.

University of Texas professor Mukul Sharma has demonstrated that strategically timing spatially contiguous fracturing operations over a few hours reduces the spatial extent of the stress shadow and, therefore, fracture interference. This method of timing of fractures ensures more efficient fracture stages without use of specialized tools and results in maximum reservoir exposure through fracturing, allowing more efficient drainage at no increase in cost.

Sharma has developed a method based on a novel downhole electromagnetic (EM) tool that allows imaging of the propped portion of induced fractures. His tool and method of EM logging provide a

way to accurately determine the distribution of proppant throughout the induced fracture network, unlike current methods such
as microseismic monitoring that only detects locations of shear
failure. This allows one to map propped fractures up to several
hundred feet away from the wellbore in both open-hole and cased-
hole completions, and provides the only way to map propped fractures at a much lower cost than existing diagnostic methods.

Credit: Carey W. King *for the UT Austin Energy Institute*

NEAR AT HAND   **University of Texas's novel membrane
reduces fracing water consumption**

University of Texas research has developed a process to reduce
the amount of fresh water used in hydraulic fracturing by recycling
up to 50 percent more water than existing techniques.

Flowback and produced water from oil and gas production are
toxic. They contain oil, salt, and an array of minerals and chemicals that must be removed before being returned to the environment or reused in further drilling. One of the biggest problems in
the oil field is what to do with this water—treat and reuse the dirty
water or inject it into disposal wells? Cost-effective membranes
for water treatment can avoid new water acquisition and the need
to dispose of produced water.

Professor Benny Freeman's state-of-the-art filtration system
developed at UT takes produced water and pumps it through two
specially coated membrane filters. The first is coarse and removes
oils, chemicals, and large contaminants like rocks. The second
uses reverse osmosis, the same technology used to desalinate
seawater, removing its salts and minerals. The result is a supply of
recycled water ready to be injected back underground to free more
rock-bound pockets of oil and gas. Typically, cleaner water yields a
more efficient fracing process, as much as 50 percent higher with
these treated membranes. Freeman perfected his membrane coatings while trying to increase the efficiency of filtration systems to
clean oily water from naval ship engines.

Their new coating, called polydopamine, is a hydrophobic coating, meaning the filter is less attracted to water, oil, and other contaminants. The filter thereby resists clogging, or "fouling." Membranes coated with polydopamine increase water throughput, thus using less energy. The membranes are also easier to clean and last up to twice as long, thus they are now in the process of being commercialized.

Credit: Carey W. King *for the UT Austin Energy Institute*

NEAR AT HAND   Stanford's ambient seismic oil field-monitoring technology

Research out of Stanford University could help oil companies improve recovery techniques and let drillers more effectively monitor existing oil fields in the North Sea.

"Many fields worldwide have shown problems in the overburden," or their ability to withstand the pressure compressing an oil reservoir, says Sascha Bussat, a Norwegain researcher for the energy company Statoil. Traditionally, oil companies monitor and manage this problem by relying on expensive and time-consuming active seismic surveys, in which boats float out into the great, stormy maritime expanse with guns full of compressed air, which blast sound miles deep into the seabed.

Stanford graduate student Sjoerd de Ridder aims to bridge the data disconnect presented by big-ticket, active seismic testing by offering a more affordable model: continuously recording fainter seismic waves naturally occurring at the ocean floor as water moves along the crust of the earth. These passive, or ambient, seismic tests could better pinpoint where and when expensive surveys should optimally be used. The value of de Ridder's ambient seismic approach lies in harvesting data in real time and for almost no additional cost above normal operations. "That makes any value equation good," he says.

BP, the operator of Norway's Valhall oil field, has been moving to drive ambient seismic testing research forward. When de Ridder began diving further into mapping Valhall's oil deposits in 2008, scientists had known about ambient seismology for decades, but they had not yet applied it to monitoring oil reserves. De Ridder spent the summer of 2012 in Norway where initial tests proved successful, and expected that, within a year, the proper infrastructure would be in place to monitor seismic signals in the North Sea 24/7.

Credit: Julia Barrero for *Stanford Peninsula Press* ("*Stanford's New Oil Mapping Technology Could Quickly Uncover More Petroleum,*" May 22, 2013)

Looking beyond the next decade, and even as the US energy system continues to greatly expand the usage of alternative energy, it is nonetheless clear both that oil and gas will maintain a major share of our energy demand and remain important pillars of our national and local economies. So longer-term R&D is just as important in this sector as it has always been.

For example, if China were able to successfully exploit its significant but hard-to-reach shale-gas resources as the United States does today, it would be just as much, or more, of a game changer as our own domestic experience to date. And China is even more desperate for solutions because of the devastating effect of particulate pollution from coal-fired power on its cities. For hydraulic fracturing to work in China, in addition to an improved property rights regime, we will need new technologies to improve deep-well costs and reduce fracing's water use. This is hard, but not impossible. Moreover, shifting natural gas from what is now essentially an imported boutique fuel in China to a serious domestically produced baseload power-generation contender would displace coal and help enable further renewable deployment on the Chinese electric grid through the ability to balance intermittence. Both would have positive spillover effects beyond China.

In that spirit, US universities are forging ahead in R&D for radical new drilling and completion concepts. Moreover, many such subsurface technologies could be dual use, enabling other clean but currently less economic energy systems such as geothermal and carbon capture and sequestration.

ON THE HORIZON   MIT's millimeter-wave directed energy drills

Accessing critical resources such as geothermal energy and natural gas drilling is an expensive, energy-intensive, messy process with today's technology. But researchers have been looking into a more elegant approach. Instead of grinding rock to bits, they would use a continuous beam of energy to vaporize it and then blow out the tiny particles that form with a high-pressure stream of injected gas. Using a device borrowed from nuclear fusion research, Paul P. Woskov, of MIT's Plasma Science and Fusion Center, has vaporized rock for the first time, confirming the feasibility of his proposal to use energy beams rather than drill bits to access underground energy resources.

For industry audiences, a key source of skepticism about the proposed approach is the absence of "drilling muds"—fluids that are carefully tailored to both remove debris and strengthen the walls by plugging up pores in the rock and providing back pressure that counters inward pressures on the hole. "Deep drilling without drilling mud is unheard of," says Woskov. However, he suggests that the glassy, or "vitrified," walls in his system could be strong enough not only to keep the hole open during drilling, but also to withstand the extreme pressures on the finished borehole with no added cement or metal liner.

"I think we have the potential to revolutionize drilling, but it's a completely new approach that will throw out a lot of conventional wisdom. That's the problem," says Woskov. He likens it to the 1940s, when the aircraft industry was working with mechanical rotary engines and propeller technologies. When the jet engine came along, it was a completely new approach that took away all the mechanics and worked on a jet stream of air. Says Woskov, "It's the same thing. Drilling technology is in the mechanical age right now, and we want to move it into the jet age"—in this case, the age of directed energy.

Credit: Nancy Stauffer, ©*Massachusetts Institute of Technology, used with permission (MIT Energy Initiative, "Reaching Underground Resources," October 16, 2012)*

ON THE HORIZON University of Texas's research into carbon dioxide storage linked to enhanced hydrocarbon production

A variety of research programs at the University of Texas are looking at the geologic storage of carbon dioxide (CO_2) in hydrocarbon-producing formations.

For example, the US Gulf Coast is well known for its oil-producing offshore rigs, but it also has unique geology that can store large quantities of CO_2 more than a mile underground in hot, salty fluid. The distinguishing feature of this brine is that it is saturated with dissolved methane. Bringing the brine to the surface with extraction wells allows for the recovery of vast amounts of geothermal energy, and once CO_2 is injected into the brine, it forces out significant amounts of methane. The CO_2-laden brine can then be sent back down for permanent storage. Recent calculations by professor Steven Bryant show that enough deep brine exists along the US Gulf Coast to store one-sixth of the country's CO_2 emissions and to meet one-sixth of its demand for natural gas annually.

Other research, led by UT research scientist Katherine Romanak at the Gulf Coast Carbon Center, has developed a new soil gas method for characterizing near-subsurface CO_2. By discriminating between CO_2 that potentially leaked from an injection location versus CO_2 generated in situ by natural processes, this method can identify CO_2 that has leaked from deep geologic storage reservoirs into the shallow subsurface. Whereas current CO_2 concentration-based methods require years of background measurements to quantify variability of naturally occurring geologic CO_2, this new approach examines chemical relationships between in situ N_2, O_2, CO_2, and CH_4 to promptly distinguish a leakage signal. The ability to measure in this way, without the need for background measurements, could decrease uncertainty in leakage detection at a low cost. Trials have already been performed as part of the investigation into CO_2 seepage from the Weyburn enhanced oil-recovery project in Canada.

Credit: Carey W. King *for the UT Austin Energy Institute*

CHAPTER 2

SOLAR PHOTOVOLTAICS

This is absolutely critical so we're hell bent on getting this done.
—Arun Majumdar, *Google vice president for energy**

Renewable solar power, which has the potential to enhance the energy security and environmental performance of the US energy system, is a reality today. The extremely low cost of the conventional baseload power generation that is otherwise available to Americans, however, means that renewable technologies—whether centralized or distributed—require a continued strong R&D effort to compete economically without subsidies.

But here is why so many observers are optimistic about photovoltaics (PV): things are getting better, and fast. Take a pencil to the historical solar PV cost per kilowatt-hour curve, and extend its gradual declining slope just five or ten years into the future. Before you know it, the pencil will start getting close to zero, and well within the cost range of many conventional power-generation technologies—all with little to no fuel requirements or operational pollution. While there are legitimate questions about the sustainability of this decline (i.e., how much of it is the result of genuine innovation and economies of scale and how much of it is a result of low-cost dumping of PV panels by a Chinese manufacturing glut), there is no doubt that trends are moving rapidly in the right direction.

If you compare the cost of solar fifteen years ago to the cost of solar today, it has been reduced dramatically by a factor of five. And we're expecting that in the future there will be another factor of three at a minimum. There is nothing that is fundamentally stopping us from getting the cost down.

—Vladimir Bulovic, *MIT School of Engineering associate dean for innovation*

Today, there is a lot around the edges that can be done to make solar PVs more affordable: the pick-and-shovel work of installation that now makes up a majority of the total cost could be reduced, permitting could be streamlined, or more efficient financing mechanisms could extend consumer availability. At the same time, as noted above, many of the PV panel-side cost reductions in recent years have been the result of fierce competition among manufacturers, slashing profit margins and optimizing operational efficiencies to make up for over-investment in production capacity while tweaking well-known panel technologies to eke out slight gains in efficiency, form factor, and panel longevity.

But ultimately, R&D in the lab amplifies the effectiveness of all of these "soft cost" efforts, which promises to significantly improve the performance of the underlying PV product. For example, even the bugbear of panel-installation costs in many ways hinges on R&D—if not through direct research on panel form-factor or attributes such as weight and modularity, then on fundamental if tedious marginal progress in panel conversion efficiency itself. After all, if the same PV panel that costs $100 to lift onto a roof is more efficient to the tune of 10 percent, then the share for installation in the total cost may go up, but the actual installation costs per kilowatt-hour of electricity produced will fall. Driving down this cost per unit of useful energy production motivates much of today's R&D efforts in PVs.

We're doing work on new substrates to get solar photovoltaics off of glass. And that speaks directly to potential reduction in weight. If we can tackle the weight problem, it will make a huge impact on installation costs, and in this area I think there are many opportunities for materials-based innovations.

—Jeffrey Grossman, *MIT professor of materials science and engineering*

PV technology based on "first-generation" crystalline silicon solar cells is the most widespread solar technology today. 86 percent of the solar cells shipped in 2011 were based on wafers of silicon. According to DOE, silicon modules with a power-conversion efficiency of 15 percent are now sold for approximately $1 per watt, the price to build the racks that support the solar cells and to install them is $1.30 per watt, and the price of the power electronics that convert the DC electricity coming out of solar cells into the AC electricity that our electricity grid uses is $0.30 per watt. So, the total price of the system is $2.60 per watt. A current overarching research goal is to reduce that total system cost to just $1 per watt (it is important to note, however, that these costs do not include firming this intermittent source of power, which adds a substantial cost to the overall system). Arriving at this target by reducing the price of the modules to $0.50 per watt, installation to $0.40 per watt, and power electronics to $0.10 per watt will require advances in logistics and permitting, improved power electronics to reduce inverter costs, lighter module substrates to reduce needed mount materials and installation equipment, and more efficient panels that

produce more electricity per any given balance-of-system resid-
ual unit.

$1 per watt installed system cost. That's a game changer. With
that, you don't need subsidies to really make solar to fly, to be
comparative to other sources of energy.

—Yi Cui, *Stanford David Filo and Jerry Yang faculty scholar,*
associate professor in materials science and engineering

**AVAILABLE TODAY   High-efficiency monocrystalline
rear-junction silicon cells**

The SunPower Corporation is a multibillion-dollar company that
recently saw a major investment from Total New Energy S.A. Its
high-efficiency solar PV cells are found on hundreds of thousands
of US households and businesses and NASA aircraft. But behind
the headlines, this Silicon Valley-based, Stanford professor-
founded clean-energy startup has actually been selling—and qui-
etly refining—its silicon wares for nearly thirty years.

Richard Swanson was a graduate student in the Stanford electrical engineering department in the 1970s when he began investigating how the PV technology then reserved for use in satellites
could be reduced from its truly atmospheric cost of over $70 per
watt. While many were looking to thin-films or other radically new
technology pathways, his "point-diffused contact" cell designs
aimed to improve efficiency by concentrating light and capturing
the current produced at the back of the cell. This increased the
amount of light available for conversion to electricity, reduced
the need for expensive wiring throughout the cell, and helped
open the door to automating manufacturing.

By the 1980s, Swanson, now running his own lab as a Stanford professor, was supported in his work by both the US DOE and Electric
Power Research Institute (EPRI), the electric power industry's
collaborative research organization. In founding SunPower,
Swanson and his team pivoted their technical approach to follow

the small but growing PV market: fabricating increasingly large wafers, thinning cells, and using wire saws to reduce manufacturing costs. Incremental improvements have actually paid off: world-record monocrystalline cell efficiencies based on Swanson's innovations have gradually risen to now exceed 24 percent. Moreover, panel costs continue to fall as production increases— what economists call "learning curves." Today, "Swanson's Law" refers to the trend of module prices falling by approximately 20 percent for every doubling of cumulative worldwide production.

As first-generation crystalline solar cells continue to be deployed and marginally improve, R&D efforts continue on a parallel path. Many researchers believe that it should be possible to make solar modules that are cheaper than those based on silicon by depositing thin-film semiconductors on inexpensive substrates made of materials like glass, metal, or plastic. The two most promising of these "second-generation" semiconductors are cadmium telluride (CdTe) and copper indium gallium diselenide (CIGS), which today have not arrived near their theoretical limits. There are many companies commercializing these two technologies, but there is a sense that more basic research needs to be done in the universities and national labs to better understand these materials so that more efficient and reliable solar cells can be made with them. The price of thin-film modules is currently below $1 per watt, although the lower conversion efficiency of thin films leads to a need for more surface area and associated installation costs. Researchers believe that thin films have the potential to yield modules with power-conversion efficiency greater than 20 percent at a price of 50 cents per watt. While numerous scientific and manufacturing challenges remain, both first- and second-generation PVs have become economical in various use cases faster than even very optimistic forecasts predicted just a few years ago.

Another type of second-generation technology is the organic solar cell—for example, with active layers made of polymers or even entirely of carbon. One advantage is that they do not use expensive semiconductors or rare elements, but their even lower efficiencies —between 1 and 10 percent in the lab—and relatively short lifetimes remain a challenge. Continued R&D is making steady progress in this area on cost reduction, higher efficiency, and greater reliability.

> We have worked on organic solar cells for a long time. These are based on molecules that can be sprayed onto a flexible plastic substrate at very low cost. When I started ten years ago, efficiency was down at a couple of percent. But this is rising very fast. And now the world record efficiency is up to eleven percent and no signs that it's slowing down. So we think we can make a lot of progress there.
>
> —Michael McGehee, *Stanford professor of materials science and engineering*

Directly addressing the balance-of-system and installation costs of PVs, robust, reliable, and high-performance panel-level "maximum power point" sun trackers or better DC/AC inverters could potentially increase overall PV system power outputs by as much as a few percentage points—a significant jump in an area that grinds away for efficiency gains measured by tenths of a percentage point. There are other novel ways of tackling the installation problem, too—for example, solar panels can be combined in the absence of sun tracking to build macroscopically three-dimensional PV structures that generate measured base-area energy densities that scale linearly with height. Other research and design efforts aim to reduce installation cost by making solar panels so easy to install and safe enough that even unskilled workers can set them up—"plug and play."

Finally, as intermittent renewable power generators increasingly populate the nation's power grid, researchers are making progress toward better handling of that intermittency at the grid level and increased generator output through improved forecasting techniques—for example, wind and cloud movements—and end-user load management.

Last year we dumped 325 gigawatt hours of wind power because it wasn't optimized to the grid. Just out of air. That's bad. We could do a better job with better-coupled grid and renewables forecasting capabilities.

—Julio Friedmann, *Lawrence Livermore National Laboratory chief energy technologist**

NEAR AT HAND   **Stanford's nanocrystalline-silicon shells improve thin solar panel light absorption**

Visitors to Statuary Hall in the US Capitol may have experienced a curious acoustic feature that allows a person to whisper softly at one side of the cavernous, half-domed room and for another on the other side to hear every syllable. Sound is whisked around the semicircular perimeter of the room almost without flaw. The phenomenon is known as a whispering gallery. A team of engineers has now created tiny hollow spheres of PV nanocrystalline-silicon and harnessed physics to do for light what whispering galleries do for sound. The results could dramatically reduce materials usage and processing cost in the production of solar panels.

"Nanocrystalline-silicon is a great PV material. It has a high electrical efficiency and is durable in the harsh sun," says Shanhui Fan, a professor of electrical engineering at Stanford. The downfall of nanocrystalline-silicon, however, has been its relatively poor absorption of light, which requires thick layering that takes a long time to manufacture.

The engineers call their spheres nanoshells. Producing the shells takes a bit of engineering magic. The researchers first create tiny balls of silica—the same material that makes glass—and coat them with a layer of silicon. They then etch away the glass center using hydrofluoric acid that does not affect the silicon, leaving behind the all-important light-sensitive shell. These shells form optical whispering galleries that capture and recirculate the light. "The light gets trapped inside the nanoshells," says Stanford's Yi Cui, an associate professor of materials science engineering. "It circulates round and round rather than passing through and this is very desirable for solar applications."

Credit: Andrew Myers *for the Stanford School of Engineering ("Stanford Engineers' Nanoshell Whispering Galleries Improve Thin Solar Panels," February 7, 2012, in the Stanford Report)*

NEAR AT HAND   MIT's vertical PV panel structures reduce installation cost

Intensive research around the world has focused on improving the performance of solar PV cells and bringing down their cost. But little attention has been paid to the best ways of arranging those cells. Now, a team of MIT researchers has come up with a new approach: building cubes or towers that extend the solar cells upward in three-dimensional configurations.

The results from the structures they have tested show power output ranging from double to more than twenty times that of fixed flat panels with the same base area. Researchers saw the biggest boosts in power in the situations where improvements are most needed: in locations far from the equator, in winter months, and on cloudier days. The new findings are based on both computer modeling and outdoor testing of real modules.

"I think this concept could become an important part of the future of PVs," says the team's leader, Jeffrey Grossman, associate professor of material science and engineering. The time is ripe for such an innovation, Grossman adds, because solar cells have become less expensive than accompanying support structures, wiring, and installation. Self-supporting 3-D shapes could even create new schemes for residential or commercial PV installation, and the increased energy density could facilitate the use of cheaper thin-film materials in area-limited applications. As the cost of the cells themselves continues to decline more quickly than these other costs, they say, the advantages of 3-D systems will grow accordingly.

Credit: David Chandler, ©*Massachusetts Institute of Technology, used with permission (MIT News Office "A New Dimension for Solar Energy," March 26, 2012)*

The DOE "Sunshot" team has just started what they call the "Michael Jordan" program. How can you get to 23 percent efficiency from silicon PV? 23 percent being Michael Jordan's uniform number. There are issues about material quality. How many impurities do you have? What's the mobility? What kind of recombinations are you getting? How do you improve the contacts? I don't think we know all the answers to that. That's a research problem that will have a direct implication on manufacturing.

—Arun Majumdar, *Google vice president for energy**

Third-generation solar technologies that aim for much higher efficiencies are also in development. "Multijunction" solar cells contain a stack of cells that harvests sunlight from different parts of the solar spectrum including both high- and low-energy photons that might otherwise be wasted. These modules, with lab efficiencies today as high as 44 percent and theoretical efficiencies nearly twice that, are currently much too expensive for large-scale deployment—$50,000 per square meter versus $70 per square meter for CdTe—but find use in niche applications such as spacecraft. The current market for this technology is therefore in pairing these solar cells with highly concentrating arrays—systems the size of an IMAX theater screen that nonetheless produce significant amounts of power. However, a continuing robust R&D program may bring the costs of multijunction cells and of other advanced solar PV concepts into line for the long term.

For example, researchers are investigating "tandem" solar cells as a step toward multijuntion's high efficiencies while still keeping costs low. This approach uses individual layers of concentrator cells, but operating under natural, single-sun conditions. Options include stacking of materials with optimal solar wavelength absorption bandgaps that are very thin and therefore flexible—with resulting efficiencies of over 30 percent.

> The approach that I'm personally getting really excited about is stacking two cells on top of each other. I believe that organic cells can be put right on top of the conventional silicon or CIGS cells.
>
> —Michael McGehee, *Stanford professor of materials science and engineering*

Other long-term solar research is looking at novel and potentially very low-cost applications that might make distributed solar power production feasible from unconventional materials more suitable to the rural developing world. One potential desirable attribute here is module flexibility. Another is extreme light weight. Both could also be beneficial to military operations when power is needed in places where it is difficult to obtain fuel.

The proliferation of solar is about scale . . . immense scale along
with the speed and cost of installation. Today, we can deploy
infrastructure at scale and speed while meeting cost constraints.
As we strive to further enhance solar's attractiveness, we need to
learn to install it as economically as we roll out other infrastructure
like roads.

—Vladimir Bulovic, *MIT School of Engineering associate dean
for innovation*

**ON THE HORIZON   MIT's thin-film organic polymer flexible
solar cells**

Using a novel process involving moderate temperatures and no
liquids, a team of researchers at MIT has printed PV cells on tissue paper, printer paper, newsprint, textiles, and even plastic food
wrap. These solar devices have features that make them ideal not
only for integrating into consumer products but also for shipping
to remote regions of the world where energy demand is growing
rapidly and there is no power grid in sight.

Today's PVs are typically fragile and must be moved with care and
installed by trained experts to avoid damage. More robust PVs have
been made on flexible materials such as plastic, but thus far, they
have not been entirely successful: problems have stemmed largely
from the anode (the positive electrode), which has a tendency to
crack or lift off when the surface on which it is mounted is bent.

But anodes that are lightweight, flexible, and adhere well have
now been fabricated by Karen K. Gleason, professor of chemical engineering, and her colleague Vladimir Bulovic, MIT School
of Engineering associate dean for innovation, with their "Paper PV
Team." Key to their success is a process called oxidative chemical
vapor deposition, or oCVD. Invented by Gleason, oCVD improves
on conventional CVD, a well-known method of depositing a thin
coating of one material on the surface of another, by adding an oxi-
dant and carefully selecting starting materials to enable "gentler"

heat and atmospheric operating conditions inside a vacuum cham-ber. The oCVD approach is designed especially for making thin films from organic polymers—carbon-containing molecules that are composed of repeating structural units and offer desirable traits including low cost, good electrical conductivity, and good mechanical properties that allow them to be flexed, stretched, and even folded.

Credit: Nancy Stauffer, ©*Massachusetts Institute of Technology, used with permission (MIT Energy Initiative "Solar Cells Printed on Paper," December 22, 2011)*

ON THE HORIZON   University of Texas's printable inorganic thin-film flexible solar cells

Novel nanocrystal inks could provide a new materials platform for creating PV devices with high efficiency on a variety of uncon-ventional substrates. Thus far, research has focused on CIGS nanocrystals as the primary light absorber material. The result-ing efficiencies of this process are good, but it requires very expensive, high-temperature sintering that is not easily scaled to large-area devices and large-manufacturing production capaci-ties, preventing commercial adoption. Nanocrystal inks, however, provide a degree of processing flexibility that cannot be obtained using processes that require high temperature.

Professor Brian Korgel's research group at the University of Texas has pioneered the synthesis of "ink-like" colloidal CIGS nanocrys-tals for PV devices. He showed that CIGS quaternary compounds could be made in high yield with good size control and dispersibil-ity using colloidal, solution-phase methods. Over time, the Korgel research team has worked to develop and improve the efficien-cies of such nanocrystal devices from less than 0.5 percent to now almost 4 percent. Moreover, this new semiconductor absorber layer can be deposited at room temperature, under ambient pres-sure, in air.

While CIGS from printed nanocrystals in Korgel's group have reached 7 to 8 percent efficiency with selenization at over 500 degrees Celsius, the team's ongoing research goal is to achieve efficiencies greater than 5 percent without the need for expensive high-temperature processing and the need for selenium vapor.

Credit: Carey W. King *for the University of Texas Austin Energy Institute*

ON THE HORIZON   University of Michigan's exploration of organic, tandem PVs

Imagine paints that can collect solar energy to power a car or living room windows that harvest the sun's rays to generate electricity for a home. A research team at the University of Michigan is analyzing the efficiency, reliability, and potential of organic PV technology for widespread commercial application.

Professor Steve Forrest's work examines the promise of capturing and distributing the power of the sun via materials that are lighter, more pliable, and less expensive than current silicon-based PV systems. "The type of organic materials we use are not very different from the inks in an inkjet printer or the dyes used in clothing," he says. "Some are very good semiconductors. In principle, they can be put down very cheaply on plastic films, metal foils, and other flexible substrates."

The research is based on proprietary small-molecule systems developed by Forrest with Global Photonic Energy Corporation, where he serves as a research partner. These systems incorporate semitransparent, organic materials stacked in a tandem architecture. This arrangement maximizes the ability to capture photons—the bundles of sunlight energy—passing through the cells. The tandem structure, Forrest estimates, achieves about 30 percent more efficiency than single-cell architecture.

While Forrest said the technology remains "next generation," its potential is far-reaching. "Commodity electricity generation is a pretty long step from here," he says. "That requires a very proven and mature technology. Still, there are a lot of interesting niches that could be exploited before that—coatings on windows and coatings on car surfaces, for example. From there, you could move into rooftop residential and then to commodity generation solar farms."

Credit: Amy Mast *for the University of Michigan Energy Institute*

John Hennessy on Game Changers
President of Stanford University

Even though there are many technologies available to us today, we as researchers still need to be thinking about energy game changers.

And despite the challenge, we do have proof that changing the game in energy technology is truly possible. I visited the University of Michigan recently and found myself looking at a classic car: a Studebaker electric auto. I did some background research on this and it turns out that in 1899 the land-speed record was broken with the first vehicle to go more than 100 miles per hour—and it was with an electric vehicle. Then there was a real game changer, called the internal combustion engine, and Henry Ford took that and changed the way we think about transportation. In a relatively quick period—ten years—we switched from electric vehicles to internal combustion engines, which was the better technology at that time.

Today again, we need to think about things that can create that kind of change—and as we think about energy "game changers," it is very important for the research community and the country to understand that we need to commit to a long, long investment horizon on energy. We need to be thinking, not just about five years, or even that something that will immediately change the game in the next fifteen years. This is important, but it is also our job as researchers to be thinking about things that are going to change the game over the next thirty or forty years.

Something that gives me tremendous hope that we are really going to get focused on this problem is not only the great young faculty we have working on it, but also the incredible energy that our students are bringing to it. They are willing to talk about new technologies and new approaches, and really dig in and commit themselves to solve it. That is what this country has that is really tremendous, and if we can bring it to bear on the problem, I think we will make a lot of progress.

CHAPTER 3

GRID-SCALE
ELECTRICITY STORAGE

Electricity is the most perishable commodity. The moment it is
generated, it needs to be consumed or stored.

—Vladimir Bulovic, *MIT School of Engineering associate dean
for innovation*

For the past century, the United States' electric sector has continuously expanded to keep up with the steadily growing demand
for electricity. The interconnectedness of the various independent systems makes the national electric system vulnerable. The
effects of an outage, either because of technical failure or terrorist attacks, would have widespread impact on the system.
What we need now is a grid that is more distributed, flexible,
and robust.

The future electric grid should enable reliable operation in
the face of uncertainty in both supply and demand. It should
integrate renewable and distributed generators. It should be fully
instrumented, supporting new power electronic components
that can enable distributed decisions and a communication network that provides data to various decisionmakers. Constant
feedback is needed, and we need a better understanding of what
needs to be monitored to manage risk. The optimization and
design of this system will require a tremendous amount of R&D
that requires modeling and simulation validated by hardware.
The incredible amounts of data that will be generated need to be
captured in a way to quickly determine what is useful, and what

is not. Furthermore, the connectivity goes beyond the physical and cyber layers to include electricity markets, enabling communication between utilities and customers so that demand can respond to real-time price signals.

A key part of all of this is the distributed large-scale storage or on-demand dispatch of electricity through fuel cells, directly in batteries, or even more exotic technologies. Our inability to cheaply store the electrons we generate at massive scale has become a major bottleneck to the performance of our country's energy system. While we have long had an efficient way to store limited amounts of electricity using pumped hydro storage whereby water is pumped uphill in massive open mountainous reservoirs, its potential for further expansion is severely limited by the availability of appropriate geography. Nonetheless, pumped hydroelectric storage represents 99 percent of worldwide storage capacity. Compressed air energy storage ranks a distant second to pumped hydro, though its more flexible siting options have reinvigorated research interest into improving this decades-old technology. Meanwhile, EPRI estimates that all battery storage technologies combined today offer an installed capacity of less than one full-size coal power plant.

And that is why research into grid-scale storage has exploded in material science, chemical engineering, and chemistry departments across the United States over the past five years. The efforts now underway draw on the best researchers from across the world with an unprecedented focus. Moreover, early achievements here are increasingly promising and coming faster than expected.

I was part of the California Council of Science and Technology's effort to see how we could get to 80% below 1990 emissions by 2050. We looked at everything; we looked at the context that the whole country was working on the same kind of problem; and we looked at biofuels and carbon capture and storage and nuclear energy and all the rest of it, and we came to the conclusion that you can't do it with renewables alone without a nonexisting technology. That nonexistent technology is what we called zero emission load: grid-scale energy storage with no emission. If you'll invent grid-scale energy storage with no emission, we saw how you could get there.

—Burton Richter, *Stanford University Paul Pigott professor emeritus in the physical sciences and SLAC National Accelerator Laboratory director emeritus*

AVAILABLE TODAY   Compressed air energy storage

Compressing air with extra energy on hand and then getting it back out when needed is not a new concept. The city of Paris used a "pneumatic dispatch of power" network in the late nineteenth century to drive small machines, water pumps, and refrigerators. More conventional utility-scale electricity storage applications followed: hundreds of megawatts of capacity in underground salt domes in Germany in the 1970s and saline caverns in the 1990s in Alabama. In the years since, however, the technology has largely sat on the shelf.

The rise of intermittent renewables such as wind and solar power has now changed that, with utilities around the country taking a second look at deploying this relatively "mature" technology. One major attraction is the cost, which is more comparable to pumped hydro storage than today's chemical batteries. Another is scale, which is potentially one hundred times that of alternatives.

"Compressed air energy storage v2.0" R&D has looked to reduce costs and improve efficiency by changing how these systems handle heat. When air is compressed, it gets extremely hot; the so-called "diabatic" systems deployed in the past lost this energy. "Isothermal" storage, on the other hand, reduces that wasted heat by spraying a fine mist into the compressing air, capturing the heated water separately for later use during expansion. A startup founded on research conducted at Dartmouth College's Thayer

School of Engineering is attempting to commercialize just this—a 1.5-megawatt unit was recently demonstrated in New Hampshire.

Another recent R&D focus has been the development of smaller, megawatt-scaled systems either above ground or in man-made caverns. Smaller, modular systems have the flexibility to be placed where they are most useful on the grid. Building on research carried out in the 1980s by Stanford graduate student Steve Chomyszak focusing on efficient "toroidal intersecting vane compressors" (initially envisioned for use in automotive hydrogen fuel cells), startup firm General Compression has recently completed its first 2-megawatt, wind turbine-driven compressed air storage facility in Texas.

AVAILABLE TODAY   USC and Caltech's direct methanol fuel cells for backup and off-grid power

When you think of backup electricity for when the grid goes down, a noisy generator probably springs to mind. An increasingly available option, however, might be fuel cells. They are clean, quiet, and have few moving parts. But compared to a gasoline or diesel generator, they are not "corner hardware store" convenient: they can be large and heavy; they need gaseous fuels such as purified methane or even hydrogen, which is hard to transport and store; and they must operate hot and at high pressures. But a type of fuel cell developed by researchers over the past two and half decades at the University of Southern California's Loker Hydrocarbon Research Institute in collaboration with Caltech's Jet Propulsion Laboratory (JPL) changes that.

With funding from DARPA beginning in 1989, USC's George Olah and Surya Prakash found that solutions of liquid methanol—an alcohol similar to ethanol fuel—could be oxidized across a catalyst to form CO_2 and water, and an electron flow in the process. Compared to hydrogen fuel cells, this "direct methanol fuel cell technology" could run at low temperatures and atmospheric pressures across a wide range of sizes. Though the efficiency of the methanol's conversion to electricity is relatively low, the high energy density inherent in the liquid fuel helps to compensate, making this technology particularly useful for long-lasting, low-power electricity delivery.

Years of additional research and refinement of ancillary systems alongside research teams at JPL and Caltech have resulted in a viable technology now used in a wide variety of applications: from portable consumer electronics to lightweight soldier-worn power supplies, and more recently off-grid or backup generator replacement for building-sized applications. Licensing their patent portfolio to manufacturers such as SFC Energy AG in 2011 has enabled the further development and commercial availability of this unique form of energy storage and delivery.

Making electricity storage technologies smaller—and moving them closer to where electricity is used, rather than where it is produced—can make them more useful and more valuable. From a research standpoint, this generally means shifting from physical to chemical storage, of which there are two main research thrusts: fuel cells and batteries. Of the two, fuel cells that produce electricity from natural gas or hydrogen have slowly evolved from science experiments to viable products on the market today.

In particular, large stationary fuel cells that can provide continuous, high-quality distributed power from multiple fuels are becoming popular with commercial users; cloud-computing data centers, for example, value them for their high mean uptime and ability to act as a very low-pollution uninterrupted power supply with a grid backup. Hundreds of megawatts have been installed, supplied by natural gas (importantly, natural gas distribution infrastructure is far more resilient than the electricity infrastructure). Over the past five years, commercially available solid oxide fuel cells have increased in capacity from 100 to 250 kilowatts while simultaneously improving conversion efficiency from the high 40 percent to now over 60 percent. But while improved manufacturing techniques and scale have helped put conventional ceramic solid oxide fuel cell technology within reach, applied R&D still aims to improve fuel cell operational lifetimes—currently a major weakness—while reducing materials and manufacturing cost.

We often think that there are logical connections between science, development, and manufacturing, in that expected order. But actually in the area of storage, there have been a number of surprises. They didn't happen because people had ideas and they worked on them and it all went in the normal sequence, but actually some of these surprises came as a result of an unexpected outside influence and some came from mistakes.

—Robert Huggins, *Stanford professor emeritus of materials science and engineering*

NEAR AT HAND   MIT's characterization of fuel cell superlattice materials

Fuel cells make electricity by combining hydrogen, or hydrocarbon fuels, with oxygen. Now, MIT researchers have unraveled the properties of a promising alternative material structure for a key component of these devices.

The new structure, a "superlattice" of two compounds interleaved at a tiny scale, could serve as one of the two electrodes in the fuel cell. The complex material, discovered about six years ago and known as LSC113/214, is composed of two oxides of the elements lanthanum, strontium, and cobalt. While one of the oxides was already known as an especially good material for such electrodes, the combination of the two is far more potent in promoting oxygen reduction than either oxide alone. Oxygen reduction is one of two main reactions in a fuel cell, and the one that has limited their overall performance—so finding improved materials for that reaction could be a key advance for fuel cells.

The key to the material's performance, explains MIT's Bilge Yildiz, associate professor of nuclear science and engineering, is the marriage of complementary qualities from its two constituents. One of the oxides allows superior conduction and transfer of electrons, while the other excels at holding onto oxygen atoms; to perform well as a fuel cell's cathode—one of its two electrodes—a material needs to have both qualities. The close proximity of the two materials in this superlattice causes them to "borrow" one another's attributes, the MIT team found. The result is a material with reactivity that exceeds that of the best materials currently used in fuel cells; as Yildiz says, "It's the best of the two worlds."

Credit: David Chandler, ©*Massachusetts Institute of Technology, used with permission (MIT News Office "Unleashing Oxygen," April 30, 2013)*

NEAR AT HAND   Stanford's use of graphene to replace platinum in fuel cells

The high price of platinum catalysts used inside fuel cells has provided a roadblock to widespread use. Now, chemistry professor Hongjie Dai's nanoscale research at Stanford University has found a way to reduce the cost.

Over the past five years, the price of platinum has ranged from just below $800 to more than $2,200 an ounce. Among the most promising low-cost alternatives to platinum is the carbon nanotube—a rolled-up sheet of pure carbon, called graphene, that is one atom thick and more than 10,000 times narrower than a human hair. Carbon nanotubes and graphene are excellent conductors of electricity and relatively inexpensive to produce. For the study, the Stanford team used multiwalled carbon nanotubes consisting of two or three concentric tubes nested together. The scientists showed that shredding the outer wall, while leaving the inner walls intact, enhances catalytic activity in nanotubes, yet does not interfere with their ability to conduct electricity.

"A typical carbon nanotube has few defects," said Stanford post-doctoral fellow Yanguang Li. "But defects are actually important to promote the formation of catalytic sites and to render the nano-tube very active for catalytic reactions." For the study, Li and his colleagues treated multiwalled nanotubes in a chemical solution. Microscopic analysis revealed that the treatment caused the outer nanotube to partially unzip and form nanosized graphene pieces that clung to the inner nanotube, which remained mostly intact.

"We found that the catalytic activity of the nanotubes is very close to platinum," says Li. "This high activity and the stability of the design make them promising candidates for fuel cells."

Credit: Mark Shwartz *for the Stanford Precourt Institute for Energy ("'Unzipped' Carbon Nanotubes Could Energize Fuel Cells, Novel Batteries," May 28, 2012, in the Stanford Report)*

But while fuel cells are becoming a more commercially viable way to supply clean electricity at small scale and on demand, they are generally designed to answer only half the distributed storage problem: conversion from the chemically stored energy (e.g., externally supplied methane natural gas or hydrogen) to electricity. Few are able to efficiently run in reverse as well—that is, taking in electricity and storing it in chemical form in order to time-shift its consumption.

Storing large amounts of electricity is such a difficult techno-logical challenge that up to now we have tended to avoid address-ing it head on with our conventional fossil fuel-based energy systems. With so many new and otherwise attractive energy technologies now requiring just this feat, however, a sense of urgency in the field has led to a blossoming of competing technology concepts that were not on the menu even just a few years ago.

An ambitious DOE research goal is to reduce the cost of large-scale, battery-based electricity storage (and reconversion) to $100 per kilowatt-hour and the best market leaders today are still in the range of many hundreds of dollars per kilowatt-hour. Important stationary storage parameters that ultimately affect cost in various deployments include the bill of materials, cycle and calendar life, high power, and safety. Game-changing technologies with the potential to meet these requirements include redox flow batteries, molten liquid metal batteries, and low-cost aqueous battery materials.

More broadly, a key framing question for the scientists and engineers working in this area is whether electrochemical storage (such as fuel cells and batteries) or some other form of storage will be optimum for different situations. For example, pumped water storage, the reigning electricity storage king, is attractive from a cost perspective—about one-tenth that of conventional lead-acid grid batteries—but available locations are few and may be far from demand or unsuitable as sites for the intermittent renewable energy production that needs it. To that end, many researchers today are experimenting with novel physical storage options such as above-ground, modularized compressed air or engineered dry gravity systems that could be more flexibly located or potentially operated at a smaller scale.

Though these technologies are still far from large-scale commercial deployment and call for a strong research effort, their existence speaks to the rapid progress in this field over the past decade.

In general, when we think about storage, the game changers will
come from three buckets. One is new materials—new chemistry
that will store more energy, at less cost, and will be easier to
produce. The second is the architecture of the storage device. That
starts a round of innovation that was not there previously. And then
the third is the issue of manufacturability and scale.

—Yet-Ming Chiang, *MIT professor of materials science and engineering*

ON THE HORIZON   MIT's liquid-metal grid storage batteries

A new system developed at MIT to lower the costs and increase
longevity of large-scale energy storage uses high-temperature batteries with liquid components that, like some novelty cocktails, naturally settle into distinct layers because of their different densities.

The three molten materials form the positive and negative poles
of the battery, as well as a layer of electrolyte—a material that
charged particles cross through as the battery is being charged
or discharged—in between. All three layers are composed of
materials that are abundant and inexpensive, explains MIT's
Donald Sadoway, professor of materials science and engineering. One promising recipe: magnesium for the negative electrode
(top layer), a salt mixture containing magnesium chloride for the
electrolyte (middle layer,) and antimony for the positive electrode
(bottom layer). The system would operate at a temperature of
700 degrees Celsius.

In this formulation, Sadoway explains, the battery delivers current
as magnesium atoms lose two electrons, becoming magnesium
ions that migrate through the electrolyte to the other electrode.
There, they reacquire two electrons and revert to ordinary magnesium atoms, which form an alloy with the antimony. To recharge,
the battery is connected to a source of electricity, which drives
magnesium out of the alloy and across the electrolyte, where it
then rejoins the negative electrode.

The inspiration for the concept came from Sadoway's earlier work
on the electrochemistry of aluminum smelting, which is conducted

in electrochemical cells that operate at similarly high temperatures. Many decades of operation have proved that such systems
can operate reliably over long periods of time at an industrial scale,
producing metal at low cost. In effect, he says, what he figured out
was "a way to run the smelter in reverse."

Credit: David Chandler, ©*Massachusetts Institute of Technology, used
with permission (MIT News Office "Liquid Batteries Could Level the Load,"
February 14, 2012)*

ON THE HORIZON   Stanford's long-life crystalline copper hexacyanoferrate battery electrode

Stanford researchers have developed part of new long-life grid-
scale battery: a new electrode that employs crystalline nanoparticles of a copper compound. In laboratory tests, the electrode
survived 40,000 cycles of charging and discharging. "At a rate of
several cycles per day, this electrode would have a good thirty years
of useful life on the electrical grid," says graduate student Colin
Wessells.

The electrode's durability derives from the atomic structure of the
crystalline copper hexacyanoferrate used to make it. The crystals
have an open framework that allows ions—electrically charged
particles with movements that en masse either charge or discharge a battery—to easily go in and out without damaging the
electrode. Most batteries fail because of accumulated damage to
an electrode's crystal structure.

Because the ions can move so freely, the electrode's cycle of
charging and discharging is extremely fast, which is important
because the power you get out of a battery is proportional to how
fast you can discharge the electrode. To maximize the benefit of
the open structure, the researchers needed to use the right-sized
ions. Too big and the ions would tend to get stuck and could damage the crystal structure when they moved in and out of the electrode. Too small and they might end up sticking to one side of the

open spaces between atoms, instead of easily passing through. The right-sized ion turned out to be hydrated potassium, a much better fit compared to other hydrated ions such as sodium and lithium.

"It fits perfectly—really, really nicely," said Stanford's Yi Cui. "Potassium will just zoom in and zoom out, so you can have an extremely high-powered battery."

Credit: Louis Bergeron *for the Stanford News Service (adapted from "Nanoparticle Electrode for Batteries Could Make Large-scale Power Storage on the Energy Grid Feasible, Say Stanford Researchers," November 23, 2011, for the Stanford Report)*

ON THE HORIZON   Stanford's simplified lithium polysulfide membrane-free flow battery

Researchers from DOE's SLAC National Accelerator Laboratory and Stanford University have developed a new flow battery with a simplified, less-expensive design that presents a potentially viable solution for large-scale production.

Today's flow batteries pump two different liquids through an inter-action chamber where dissolved molecules undergo chemical reactions that store or give up energy. The chamber contains a membrane that only allows ions not involved in reactions to pass between the liquids while keeping the active ions physically sepa-rated. This battery design has two major drawbacks: the high cost of liquids containing rare materials such as vanadium—especially in the huge quantities needed for grid storage—and the mem-brane, which is also expensive and requires frequent maintenance.

The new Stanford/SLAC battery design uses only one stream of molecules and does not need a membrane at all. Its molecules mostly consist of the relatively inexpensive elements lithium and sulfur, which interact with a piece of lithium metal coated with a barrier that permits electrons to pass without degrading the metal. When discharging, the molecules, called lithium polysulfides, absorb lithium ions; when charging, they lose them back into the liquid. The entire molecular stream is dissolved in an organic solvent, which does not have the corrosion issues of water-based flow batteries.

"In initial lab tests, the new battery also retained excellent energy-storage performance through more than 2,000 charges and discharges, equivalent to more than 5.5 years of daily cycles," says research group leader Yi Cui.

Credit: Andy Freeberg *for the SLAC National Accelerator Laboratory ("New Battery Design Could Help Solar and Wind Energy Power the Grid," April 24, 2013)*

Susan Hockfield on Game Changers
President emerita of MIT

For the United States, transitioning to a sustainable energy future is a great avenue for economic growth. It is critical for national security, too: in our country's history, we have come to rely on undependable countries for oil and gas; will we now rely on undependable countries as we import energy technology from them? And, of course, the urgency of climate change puts an enormously high premium on acting as quickly as possible.

One of the things that is at the heart of MITEI is industry partners. When we set up the initiative, we counted on our industry partners having special knowledge about the energy marketplace; that knowledge has proven absolutely invaluable as we design research projects to meet that marketplace. The cross-talk, the sharing of information, the acceleration of development of technologies—I do not believe we could have achieved this in any way aside from having our industry partners close to us.

But the most important thing though that we have learned from MITEI—our secret and most powerful weapon—is the enthusiasm of young people today. The students at MIT are nothing less than lunatics about a sustainable energy future. The incredible ambition of our students, their creativity, and their commitment are resources that I fear we are drawing on insufficiently. The young people in this generation—I am a little embarrassed to say—are different from the young people of my generation. The kind of change we imagined we could effect was mostly in theory and only a little bit in practice. But these young people today are really committed to making change in a real way in the real world. So for us all as energy future leaders, I think we have a big responsibility to the next generation to begin to draw out the plans for where the avenues and highways are going to go, because we have an army coming along who will be happy to march along them and make many of our ideas reality.

CHAPTER 4

ELECTRIC CARS

Moderating US oil consumption remains a high priority for security, economic, and environmental reasons. 70 percent of US oil consumption is used for transport, and the International Energy Agency estimates that the US transport sector's per capita liquid fuel consumption is the highest in the Organization for Economic Cooperation and Development (OECD), easily exceeding second-place Canada. In fact, 27 percent of all global transport energy use occurs in the United States. At the same time, mobility is a key enabler of the country's economy and so numerous R&D efforts in universities and industry are aimed at displacing the role of oil in this sector.

Students and faculty at universities around the country are devoting their efforts to finding innovative solutions for finding affordable, cleaner, more flexible, and resilient transportation options. Mechanical engineers, chemical engineers, electrical engineers, materials scientists, biologists, chemists, physicists—all working together, employing and pushing the frontiers of a whole new generation of scientific inquiry based on the idea of atomic and

molecular level design, synthesis and characterization of materials
and processes for energy conversion and supply.

—Sally Benson, *Stanford Global Climate and Energy Project director
and professor of energy resources engineering*

Using less oil—or none at all—to get where we want to go
is certainly not a new idea, but the options to do so today are
better and the necessary tradeoffs are smaller than ever before.
On the one hand, passenger vehicles are undergoing significant
efficiency improvements, and this is expected to continue over
the next decade as fuel economy requirements increase. More
efficient internal combustion engines are one part of the story,
for example the development of smaller-displacement turbo-
charged engines coupled to more versatile transmissions. This
is complemented by alternative drivetrains—conventional
gasoline-electric hybrids in particular—and vehicle envelope
improvements, such as light-weighting and aerodynamic design.
Our increasingly efficient vehicle fleet already helps to insulate
US drivers from fluctuating oil prices and improves the amount
of personal mobility available per dollar spent. And local emissions have never been better—today's "super ultra low emission
vehicles" with modern gasoline drivetrains emit roughly 90 percent less pollution than the average new vehicle sold in 2003.

If you look at diesels, their efficiencies are in the low forties now.
Actually, there are things coming out of what's called "homogenous
charge compression ignition," or HCCI—the next generation beyond
HCCI—which have indicated efficiencies approaching 55 percent
now. And those technologies will fit into relatively conventional
piston-cylinder geometries—very highly turbo charged. . . .
Efficiencies are way up, on the move.

—Chris Edwards, *Stanford professor of mechanical engineering*

AVAILABLE TODAY   MIT's carbon nanotube enhanced-ultracapacitor

Ultracapacitors are a promising energy storage device. They deliver energy quickly, can be recharged in seconds, and have a long life span—but their energy-storage capacity is limited. In 2004, MIT's Joel Schindall, B. Gordon professor of the practice, electrical engineering and computer science, proposed a solution to that problem: instead of using activated charcoal, coat the ultracapacitor electrodes with vertically aligned carbon nanotubes. A tightly packed array of tall, thin nanotubes on the electrode could provide lots of surface area for critical charged particles called ions to cling to. The nanotube "forest" would provide straight pathways so the ions could come in and out easily and pack together neatly—like sucking up paint with a paintbrush rather than a sponge.

Early work at MIT on prototype devices showed that the new ultracapacitor could store energy, but the demonstration devices were each the size of a thumbnail and could charge and discharge only tiny amounts of energy. After years of refining the technology, in 2008, Riccardo Signorelli, MIT PhD '09, cofounded FastCAP Systems, a company aimed at commercializing the nanotube-enhanced capacitor along with systems to enable its practical implementation. The company has twenty-five employees and recently sold and shipped its first generation of products.

Their latest ultracapacitor stores twice as much energy as conventional alternatives and delivers seven to fifteen times more power. It also costs less and uses raw materials that are both inexpensive and abundant within the United States. While the new ultracapacitor has potential applications in many fields, the immediate focus is on transportation. For example, in a hybrid-electric vehicle, an ultracapacitor could provide power for rapid acceleration and deceleration and instant discharging and charging—a million or more times over the lifetime of the vehicle. "Most people don't associate the word 'hybrid' with a high-performance vehicle, but our ultracapacitors could change that," says Signorelli.

Credit: Nancy Stauffer, ©*Massachusetts Institute of Technology, used with permission (MIT Energy Initiative "A Novel Ultracapacitor," October 9, 2012)*

AVAILABLE TODAY   Lithium-ion batteries in mobile electronics and plug-in electric vehicles

Lithium-ion batteries are the workhorses of modern consumer electronics. You probably have two or three within arm's reach right now. And they have found surprising early uptake in plug-in or even full-electric vehicles such as the Tesla Model S, which uses thousands of them in massive under-floor battery packs. The lithium-ion battery's ubiquity today is the result of early breakthroughs in industry and academia, followed by decades of gradual refinements on commercial production lines.

Their origin can be traced to 1974, when Stan Whittingham traded his postdoctorate research role in professor Robert Huggins's material science lab at Stanford for one at the Exxon Corporation. There, working with a research team interested in superconductivity, Whittingham found that they could effectively shuttle miniscule lithium ions carrying a small electric charge in and out of a structured titanium sulfide electrode. The result of this process, called intercalation, was one of the first demonstrated modern rechargeable lithium batteries.

Later, in 1980, John Goodenough at Oxford observed that lithium ions could be removed from a compound of lithium cobalt oxide. Concurrently, a research team at the Grenoble Institute of Technology and related work at Bell Labs showed that lithium could be inserted into layers of carbon. Together, these two breakthroughs—a positive lithium cobalt oxide electrode and a negative carbon-based anode—while essentially reversing the Exxon design, suggested a new, coherent battery architecture.

By 1985, Akira Yoshino of Japan's Asahi Kasei Corporation used these concepts to demonstrate a feasible (and safe) commercial prototype. And in 1990, Sony revealed their own surprise: a consumer camcorder powered by a lithium-ion battery. These were the first to achieve manufacturability, which would gradually pull down the price, and improve the performance, over the decades to come.

So beyond simply improving fuel efficiency, we are now seeing advances in the technological performance and commercialization of full-electric cars that would have seemed impossible just a decade ago. Lithium-ion batteries in particular have been at the top of the government's research agenda. Conveniently, many of the advances in that technology can be shared with the high-volume, high-value consumer electronics, such as iPhones. Costs have come down significantly, but still have quite a way to go for large-scale penetration of the automotive market for vehicles with long-range driving without recharge.

For example, numerous full-electric or plug-in electric sedans now compete directly with gasoline-powered alternatives at different price points. One major advantage of full-electric drivetrains is their various mechanical simplicities compared to conventional gasoline vehicles, including far fewer moving parts. But because so many batteries are needed to power vehicles over a reasonable range—say, anywhere from 25 to 85 kilowatt-hours for a range of 80 to 300 miles—electric vehicles on the market today using even the best cost-performance battery options still command an upfront price premium of about $15,000 to $20,000 before federal or state subsidies. And while cutting-edge battery designs continue to evolve quickly in the lab, keeping up with the pace of that technological change is a major challenge to upstart cell manufacturers looking to profitably commercialize such products quickly and at scale. Improved mobile storage technology is therefore key to the successful electrification of transportation—in particular, higher specific energy or energy density batteries to increase the range of vehicles and reduce costs.

Over the last ten years, lithium-ion batteries went from a portable device technology, to not only a transportation technology, but also a grid technology. That was all driven by R&D. At the same time, the price of lithium-ion was reduced from $2,000 a kilowatt-hour

to $300, which was also driven by R&D. So what are the goals?
A 3x increase in energy density along with a 3x decrease in costs.
That is game changing—at that point, it will become widely
adoptable.

—Yet-Ming Chiang, *MIT professor of materials science and engineering*

Relatively little attention was paid to vehicle batteries specifically, especially at the basic science level, until about ten years ago, and today's researchers report that significant funding did not start flowing into related R&D until about five years ago. Since then, contrary to the pronouncements of skeptics, progress on mobile battery technology has actually been very good. Of course, it may not have kept up with the breathtaking pace of microchip processing power advancements, but Moore's Law is a heady standard by any measure. For example, in just the past few years, energy density in commercially deployed vehicle batteries has improved from approximately 100 watt-hours per kilogram in the A123-type batteries shipped in the Fisker Karma to approximately 200 watt-hours per kilogram for those used in the Chevy Volt, and even breaking 250 watt-hours in the Tesla Model S. Startup companies are now testing mobile batteries at ranges of 300–400 watt-hours per kilogram and over 1000 watt-hours per liter of volume. Battery manufacturing costs for the Chevy Volt were approximately $1,000 per kilowatt-hour capacity in 2008; manufacturing costs for the same battery had fallen to $500 per kilowatt-hour in 2012. Tesla Motors now claims costs of closer to $300 per kilowatt-hour for its simplified battery pack designs, with the generic lithium-ion cells supplied by Panasonic. Cost reductions in the deployment of such vehicle batteries have significantly benefited from the manufacturing scale offered by the existing mature market for similar consumer electronic batteries.

A change in the processing can make a profound difference in the properties of the materials, the properties of the devices, and also the way in which they're used. Sony in the 1990s, with the introduction of the lithium-ion battery, disrupted the industry by doubling energy density and fundamentally changing the way they were manufactured. A game-changing battery technology both in consumer electronics and auto and later in grid storage will require rethinking manufacturing platforms—that is certainly the case with our solid-state technology.

—Ann Marie Sastry, *Sakti3 president and CEO*

Current research goals are to improve mobile battery energy density by two to five times, reduce costs per kilowatt-hour by three to five times, and improve cycle and calendar life while ensuring safety. Improvements of this scale represent a substantial research challenge. Potential game-changing technologies include the development of electrode materials to replace the graphite generally used today in order to increase storage capacity. Options include high-capacity nanosilicon anodes and sulfur, or even air-based, cathodes. Early advances in the lab suggest that combining these approaches, while effectively stabilizing the chemical degradation that can go alongside them, may yield capacity improvements of up to four times over existing lithium-ion technologies. Meanwhile, some researchers expect that entirely different chemistries will be needed to meet the necessary price points. Overall, a strong research focus on mobile storage is warranted, for both novel materials (such as nanostructured and solid-state materials) and novel manufacturing approaches (including synthetic routes to materials fabrication and fabrication of devices at scale).

Five years ago, we started work on silicon battery chemistries and they kept saying silicon doesn't work. It's not possible. After five years of continuous research and pumping a lot of ideas into this area, its looks like it has become possible. When I look at this, I get very excited because it's so difficult to solve.

—Yi Cui, *Stanford David Filo and Jerry Yang faculty scholar, associate professor in materials science and engineering*

NEAR AT HAND   Stanford's silicon nanoparticle and hydrogel lithium ion batteries

Stanford University scientists have dramatically improved the performance of lithium-ion batteries by creating novel electrodes made of silicon and conducting polymer hydrogel, a spongy material similar to that used in contact lenses and other household products. "We've been trying to develop silicon-based electrodes for high-capacity lithium-ion batteries for several years," says Stanford associate professor Yi Cui. "Silicon has ten times the charge storage capacity of carbon, the conventional material used in lithium-ion electrodes. The problem is that silicon expands and breaks."

Studies have shown that silicon particles can undergo a 400 percent volume expansion when combined with lithium. When the battery is charged or discharged, the bloated particles tend to fracture and lose electrical contact. To overcome these technical constraints, a Stanford team used a fabrication technique called in situ synthesis polymerization that coats the silicon nanoparticles within the conducting hydrogel. Using a scanning electron microscope, the scientists discovered that the porous hydrogel matrix is riddled with empty spaces that allow the silicon nanoparticles to expand when lithium is inserted. This matrix also forms a three-dimensional network that creates an electronically conducting pathway during charging and discharging.

Although a number of technical issues remain, Cui is optimistic about potential commercial applications of the new technique to create electrodes made of silicon and other materials. "The electrode fabrication process used in the study is compatible with existing battery manufacturing technology," he said. "Silicon and hydrogel are also inexpensive and widely available. These factors could allow high-performance silicon-composite electrodes to be scaled up for manufacturing the next generation of lithium-ion batteries. It's a very simple approach that's led to a very powerful result."

Credit: Mark Shwartz *for the Stanford Precourt Institute for Energy ("Stanford Scientists Create Novel Silicon Electrodes That Improve Lithium-ion Batteries," June 4, 2013)*

NEAR AT HAND   UT Austin's long-life carbon paper lithium-sulfur cells

The current lithium-ion technology based on lithium-insertion compound electrodes is limited in energy power and density. Sulfur offers an order of magnitude higher energy-storage capacity than the currently used lithium-insertion compound cathodes. Unfortunately, several challenges prevent the use of sulfur cathodes in practical cells. These challenges include low electrochemical utilization of sulfur and poor cycle life. The low electrochemical utilization of sulfur is due to the high resistivity of sulfur and the discharged products. The poor cycle life of sulfur cathodes is due to the dissolution of sulfur as intermediate polysulfides into the electrolyte and the consequent poisoning of the metallic lithium anode by the dissolved polysulfide species.

Professor Arumugam Manthiram at the University of Texas has demonstrated a novel cell configuration that overcomes the challenges and offers long-life lithium-sulfur cells with high energy and power. The new cell configuration involves the insertion of an electrolyte-permeable carbon paper between the separator and the regular sulfur cathode. The carbon paper serves as a bifunctional interlayer: Acting as a second current collector, it decreases the resistance of the sulfur electrode and facilitates the absorption and retention of the soluble polysulfide species. This novel approach not only simplifies the processing without requiring elaborate synthesis of composites and surface chemistry modification, but also enhances the charge-discharge rate by facilitating fast electron flow. With the carbon inter-layer, lithium-sulfur cells exhibit high energy with a long cycle life. The new cell configuration with the carbon paper interlayer has the potential to increase the energy density of the packaged cells by two to three times compared to the current lithium-ion technology.

Credit: Carey W. King *for the University of Texas Austin Energy Institute*

NEAR AT HAND   MIT's carbon nanofiber high-density lithium-air battery

MIT researchers have found a way to improve the energy density of lithium-air (or lithium-oxygen) batteries, producing a device that could potentially pack several times more energy per pound than today's ubiquitous lithium-ion batteries.

The team created a carbon fiber-based electrode that is substantially more porous than other carbon electrodes. It can therefore more efficiently capture the oxygen from air flowing through the system and store it as solid oxidized lithium within its pores as the battery discharges.

"We grow vertically aligned arrays of carbon nanofibers using a chemical vapor deposition process. These carpet-like arrays provide a highly conductive, low-density scaffold for energy storage," explains former MIT graduate student Robert Mitchell. The gravimetric energy stored by these electrodes—the amount of power they can store for a given weight—"is among the highest values reported to date, which shows that tuning the carbon structure is a promising route for increasing the energy density of lithium-air batteries," says former MIT graduate student Betar Gallant. The result is an electrode that can store four times as much energy for its weight as present lithium-ion battery electrodes.

In a paper published in 2010, the team had estimated the kinds of improvement in gravimetric efficiency that might be achieved with lithium-air batteries; this new work "realizes this gravimetric gain," says MIT mechanical engineering professor Yang Shao-Horn. Further work is still needed to translate these basic laboratory advances into a practical commercial product, she cautions.

Credit: David Chandler, ©*Massachusetts Institute of Technology, used with permission (MIT News Office "Research Update: Improving Batteries' Energy Storage," July 25, 2011)*

NEAR AT HAND   Berkeley's rechargeable, flexible zinc batteries

A University of California, Berkeley graduate student has resurrected a largely ignored battery chemistry in an effort to create a new class of low-cost and safe energy storage devices that could be an alternative to ubiquitous lithium-ion technology.

Zinc batteries are nothing new to material scientists—think off-brand disposable AAs—but Christine Ho's breakthrough in electrolyte design prevents formation of undesirable dendrites on the battery's zinc electrode. In conventional zinc chemistries, this buildup is so severe as to limit their use to a single discharge. In Ho's battery, the liquid electrolyte is replaced with a plastic-like yet electrically conductive gel polymer film that shuttles ions between its zinc and manganese dioxide electrodes more than 200 times. In short: this is a rechargeable device that could give zinc batteries a life beyond the trash can.

Though the technology's energy density is lower than some leading-edge lithium-ion chemistries—130 watt-hours per kilogram in early lab estimates—it has other tricks up its sleeve. Just 100 microns thick, the zinc cell is flexible. Nonreactive and nonvolatile—that is, safer—it could conceivably be integrated into final products without bulky protective packaging. And the battery is also printable, which could keep down manufacturing and material cost at scale.

While completing her doctorate, Ho joined with classmates to showcase their zinc technology in university business plan competitions. Their winnings provided enough for capital for basic prototype demonstration and the team has since formed a startup they call Imprint Energy to further push development of their innovations. Initial target markets are consumer electronics and biomedical, though the concept could be applied more broadly as well, for example, in automobiles.

NEAR AT HAND   University of Michigan's vanadium nitride supercapacitors

University of Michigan researcher Levi Thompson and his team are pairing a traditional technology with a modern workaround— the humble lead-acid battery commonly found in golf carts and electric bikes (and yes, in traditional cars) paired with a next-generation supercapacitor.

Batteries work like a watering can—they can hold a lot of energy, but they cannot pour it out quickly. By comparison, a supercapacitor is like a water tumbler—it can fill or empty much more quickly than a battery. According to Thompson, a professor in the University of Michigan Department of Chemical Engineering, a next-generation supercapacitor paired with a lead-acid battery could equal a tenfold cost savings on electric cars for automakers and consumers as compared with the lithium-ion batteries currently in use. Through his company Inmatech, Thompson aims to double the energy density of the current generation of supercapacitors while dramatically reducing the cost. The group has recently completed a five-inch cubic prototype device packed with vanadium nitride—a compound with lots of structural surface area ideal for energy storage. Inmatech's six employees will soon test the effectiveness of their invention by partnering with manufacturers of electric bikes and scooters before exploring the manufacture of automotive versions.

Other commercial applications for the technology could include memory backup in mobile phones and use in uninterruptible power supplies. There is also potential in smart-grid energy storage, as

well as in extended range vehicles and electromagnetic armor for
military purposes.

Credit: Amy Mast *for the University of Michigan Energy Institute*

Into the future, revolutions in vehicle automation and net-
working hold the promise to transform freight and personal
road transport, reducing both congestion and energy use for
whatever fuel system is used. Variations on advanced vehicle
sensing of the environment and automated technologies range
from incremental additions to driver abilities—such as adaptive
cruise control, stop-and-go traffic management, or lane depar-
ture control—all the way to full vehicle autonomy.

Most compellingly, many such technologies could both
reduce overall vehicle energy use and offer other more immediate consumer-facing benefits. For example, automated parking capability within an urban area would save time and fuel
while reducing curb-side pollution and even cutting down on
the significant urban road congestion caused by drivers circling
for available parking spaces. Similarly, to the extent that active
safety automation reduces the number of vehicle crashes, manufacturers could rely less on passive safety mechanisms such as
airbags or steel safety cages—in effect, making vehicles much
lighter. This could both dramatically improve vehicle energy
efficiency and over time mitigate the ongoing global rise in
highway mortality; already 1.3 million people are killed each
year in road accidents and an additional 20 to 50 million people
are injured according to the World Health Organization. Some
of these technologies have grown out of military-sponsored
university research—for example, the well-known "DARPA

challenges"—and are in early testing stages by vehicle manufacturers and third-party parts suppliers. Considerable work
remains, though, in proving this technology's flexibility and
safety, along with establishing both legal frameworks and user
norms.

ON THE HORIZON   **Stanford's improved highway-vehicle
induction charge system**

A Stanford University research team has designed a high-efficiency
charging system that uses magnetic fields to wirelessly transmit
large electric currents between metal coils placed several feet
apart. The long-term goal of the research is to develop an all-
electric highway that wirelessly charges cars and trucks as they
cruise down the road.

The wireless power transfer is based on a technology called magnetic resonance coupling. Two copper coils are tuned to resonate
at the same natural frequency—like two wine glasses that vibrate
when a specific note is sung. The coils are placed a few feet apart.
One coil is connected to an electric current, which generates a
magnetic field that causes the second coil to resonate. This magnetic resonance results in the invisible transfer of electric energy
through the air from the first coil to the receiving coil.

In 2007, MIT physics professors Marin Soljacic, John Joannopoulos,
and Peter Fisher used magnetic resonance to light a 60-watt bulb.
The experiment demonstrated that power could be transferred
between two stationary coils about six feet apart, even when
humans and other obstacles are placed in between. Stanford professor Shanhui Fan and his colleagues wondered if the MIT system
could be modified to transfer 10 kilowatts of electric power over
a distance of 6.5 feet—enough to charge a car moving at highway speeds. The car battery would provide an additional boost for
acceleration or uphill driving.

Fan and his colleagues recently filed a patent application for their
wireless system. The next step is to test it in the laboratory and

eventually try it out in real driving conditions. "You can very reliably use these computer simulations to predict how a real device would behave," says Fan.

Credit: Mark Shwartz *for the Stanford Precourt Institute for Energy ("Wireless Power Could Revolutionize Highway Transportation, Stanford Researchers Say," February 1, 2012, in the Stanford Report)*

ON THE HORIZON **University of Michigan's driverless vehicle test bed**

By 2021, Ann Arbor, Michigan, could become the first US city with a shared fleet of networked, driverless vehicles. That is the goal of the Mobility Transformation Center (MTC), a cross-campus University of Michigan initiative that also involves government and industry representatives.

"We want to demonstrate fully driverless vehicles operating within the whole infrastructure of the city within an eight-year timeline and to show that these can be safe, effective, and commercially successful," says University of Michigan's Peter Sweatman. The center involves researchers from all over the university, including urban planning, energy technology, information technology, policy, and social sciences.

Occupying 30 acres, the MTC is constructing a test environment that will include approximately three lane-miles of roads with intersections, traffic signs and signals, sidewalks, benches, simulated buildings, street lights, and obstacles such as construction barriers. "Connected and automated vehicles provide a new platform for safety improvements, better traffic movement, emissions reduction, energy conservation, and maximized transportation accessibility," Sweatman says. "The new facility will help the MTC accelerate and integrate innovations that will lead to a commercially viable automated mobility system that will fundamentally transform mobility in our society."

U-M has also recently implemented the world's largest on-road vehicle-to-vehicle and vehicle-to-infrastructure model deployment in Ann Arbor, with more than 3,000 users. This project, which includes several industry participants, provides data to inform future policy decisions by the US Department of Transportation.

Credit: Amy Mast *for the University of Michigan Energy Institute*

CHAPTER 5

LED LIGHTING

Technologies that give you greater efficiencies—efficiencies in lighting, air conditioning, heating, and transportation—all have the potential to decrease the cost of doing business for US companies.

—William Perry, *Stanford University professor emeritus*

Since 70 percent of electricity is consumed in our residential and commercial buildings, an energy-efficient built environment is crucial for maximizing the value of our supply-side energy infrastructure. When we talk about energy, ultimately what we really want are energy services—the proverbial cold beers and hot showers—not the kilowatt-hours themselves. So, if we can double the number of cold beers we get from one coal-fired power plant though better end use energy productivity instead of doubling the number of power plants we build, then without even touching the supply-side infrastructure, we get a more affordable and more environmentally responsible energy system.

A seminal 2009 study by the consulting firm McKinsey highlighted the massive potential cost savings—roughly $1.2 trillion over a decade—available to the United States through improved end-user energy efficiency in areas such as the building envelope, lighting, and appliances. While there are legitimate and subtle arguments about the true scale of the "efficiency gap" and associated "rebound effects" that might reduce it, there is still widespread agreement that the efficiency potential is great, that much of it remains unrealized, and that the time scale for realizing significant gains can be relatively near term.

Consumer appliance and industrial equipment efficiency are examples of areas that have made particularly good progress since the energy crises in the 1970s. A combination of energy efficiency policy and labeling programs—and of course energy prices themselves—helped create a market for technologies that have dramatically reduced the power consumption of white goods, consumer electronics, and lighting systems even as performance demands have increased. Many such technologies are available—or even dominate—the market today and have become default product choices.

A recent example worth championing: the light-emitting diode (LED) that requires only 20–25 percent of the electricity as compared to an incandescent bulb. Though first patents for LEDs were filed in the 1960s amidst work from numerous US corporate and university research groups and the technology gradually found commercial applications as indicator lighting, advances through the 1990s significantly improved LED color production and brightness enough to broaden their use in consumer goods. Shuji Nakamura of Japan's Nichia Corporation, in particular, is credited with a breakthrough demonstration of the first high-powered blue LED in 1994. Through continued university and private-sector R&D into the chemical composition of the LED, energy efficiency and brightness have approximately doubled at a pace of every three years. Costs have also come down. Even though the upfront purchase prices for "drop-in" LEDs remain high—if falling rapidly—life-cycle costs over LEDs' 25x incandescent bulb lifetime are already attractive, especially in commercial-sector general lighting applications in which such savings can be realized sooner.

Population and electricity use do not correlate. There are about 1.4 billion people who do not have access to electricity, and another 1.6 who have very limited access. So about three billion people have very limited access to electricity, or no electricity. And we're adding three more billion people within the century in most of those regions. If they start using electricity, what happens then. . . . This is a major issue, because that's where the growth is going to be.

—Arun Majumdar, *Google vice president for energy**

AVAILABLE TODAY   University of Michigan's high-quality fluorescent replacements

A University of Michigan clean energy startup looks to inform investors and future customers about a new lighting technology that offers a cleaner, longer-lasting, and higher-quality alternative to today's fluorescent tubes.

Headquartered in the U-M Tech Transfer's Venture Accelerator, Arborlight LLC has developed an LED-based, drop-in replacement for the linear, fluorescent tubes commonly used in overhead lighting and other commercial applications. These replacements are mercury free, last more than 50,000 hours, and provide a cost-effective source of uniform, bright light.

"Our current designs suggest that our lamps will be considerably more efficient, more durable, and robust than today's glass fluorescent tubes," said Max Shtein, a U-M materials science and engineering professor, who with U-M electrical engineering and computer science professor P. C. Ku conceived of the lighting architecture at the heart of Arborlight. "In addition to this being a commercial opportunity, we could improve energy efficiency in lighting and eliminate over five metric tons of mercury from the waste processing stream each year in the US alone."

Credit: Amy Mast *for the University of Michigan Energy Institute*

AVAILABLE TODAY   Affordable, off-grid LED indoor lighting

Outside of the United States, where perhaps one billion people globally rely on kerosene or other direct biomass combustion for lighting, cheap long-lasting battery-powered LEDs are already expanding access to indoor illumination, which the United Nations Foundation estimates to increase family income by about 15 percent. Moreover, this illumination comes without contributing to indoor air pollution from the combustion of solid fuels, which, according to the World Health Organization, is otherwise responsible for perhaps two million deaths worldwide each year—more than 3 percent of all global mortality.

Encouraging wider deployment of lighting is both an engineering and a design question. On the one hand, LEDs that are long lasting, high intensity, and very low power draw have come from decades of basic and applied materials science laboratory research. These are game-changing product attributes. But translating those engineering successes into compelling, problem-solving applications is not automatic.

This challenge is at the core research ongoing at the Stanford mechanical engineering department's "D.School." Here, faculty and graduate students are developing cross-disciplinary frameworks that can be used to integrate engineering, business, and behavioral science through a user-centric process they call "design thinking."

One recent spinoff: an affordable, solar-powered LED lantern for the developing world called "d.light." The four cofounders were students together at Stanford. Their experiences pointed both to the need for safe and clean indoor lighting in poor countries and to the potential for newly available LED technology to help. Years of prototyping and user testing followed: designing a usable LED and battery package, improving durability and reliability, and testing the distribution, manufacturing, and financing models needed to get their devices into the right hands. Three million lamps have since been sold to customers in Africa and India.

Like any new consumer-facing general illumination technology, however, LEDs may be desirable from an energy-efficiency standpoint, but they still have to go up against high-performance alternatives. The experience of compact fluorescent lamps (CFLs) is illustrative. There, a technology derived from industrial and commercial-sector overhead lighting applications had to be extensively tailored to be attractive to residential sector buyers. Long fluorescent tubes with external ballasts were reimagined to fit into existing incandescent bulb-style sockets, and preferably had to look good while doing it. These new CFLs had to improve heat tolerance, reduce warm-up and power-on response times, and increase their color rendering and overall light quality, all while reducing price. Ultimately, CFLs were able to displace a significant amount of energy-wasting incandescents across the United States. The adoption of LEDs has a similar R&D and commercialization roadmap today. The point is that apart from their high-energy use, incandescent bulbs are a great product for consumer-facing general lighting—they look good, they produce high-quality dimmable light instantaneously, and they are extremely cheap—so any challenger must be up to the task.

In this regard, LEDs have a number of innate attributes in their favor that CFLs never did. To start, they do not suffer from power-on delays or warm-up periods. They are dimmable and can even offer different colors. They are smaller and so reasonably suited to packaging into existing fixtures (though heat can remain a problem for ballast lifetimes). And they are more energy efficient. Ongoing R&D has focused on addressing areas for further improvement—for example, producing a high-quality and eye-pleasing white light, or improving absolute brightness levels to at least a 75-watt conventional bulb equivalent—with the goal of making LEDs not only more energy efficient, but actually achieving higher performance than incandescents. The market penetration of LEDs is about 6 percent today—which is a foothold, but with plenty of room to grow. Considering that about 13 percent of total US electricity use is for lighting in residential and commercial buildings, the energy, greenhouse gas emission, and consumer-expenditure savings of getting LEDs right could be substantial in the near to mid term.

> The potential of organic LEDs, similar to organic solar cells, is now only beginning to be fully realized because up until now we have been struggling to improve their lifetime. . . . When we made the first ones in 1980s, they would last an hour. Then we learned how to package them a little better and they would last a month. Today they last a million hours, which is 100 years of continuous use.
>
> —Vladimir Bulovic, *MIT School of Engineering associate dean for innovation*

LEDs are also playing an increasingly important role not just for general lighting but also in underpinning the television, computer, and mobile displays into which Americans spend much of their days gazing and that account for 2 percent of our total electricity use. Simply backlighting such displays with LEDs rather than fluorescent tubes reduces their energy use and heat output while increasing lifetimes. A step further is the "organic" LED (OLED) display, which is not only extremely low power (less then 0.5 watt per square inch), but also extremely thin or even flexible. OLED displays were just a few years ago the domain of multi-thousand dollar, miniscule-sized, technology demonstrations but have now found their way into broadly sold smart phone displays and other consumer electronics. OLEDs exemplify large-area nanofabrication in a consumer-facing product and account for about 4 percent of revenues in the flat-panel display market today.

NEAR AT HAND   MIT's nanophosphor "Q"LEDs

Challengers to the technological descendants of Thomas Edison's incandescent filament light bulb, including halogens, fluorescent tubes, and compact fluorescents, have time and again stepped into the ring. Each promised one major improvement—less energy use per unit of emitted light—but none has successfully delivered the knockout punch. A major weakness? Color production. But today, "quantum dots" that have been commercialized after multiyear research from the lab of MIT professor Vladimir Bulovic are set to make LEDs—Edison's latest challenger—into a true contender.

LEDs are a type of solid-state lighting—essentially, microchips—that emit a clean light with little energy lost to heat. But even after decades of improvement in their lifetime and performance, LEDs have remained impractical for general indoor illumination in part because of the cold, bluish white light they emit. This is because a common LED is actually blue—it only turns "white" to our eye when the blue photons it emits excite a phosphor coating applied around the bulb itself. The result of this coating trick? Some blue light, and some yellow light.

With Bulovic's "Q"LED quantum dots, tiny nanophosphor crystals of cadmium selenide again coat the convention blue LED bulb. But tuning the dots' diameters optically down-converts its light: 1.7 nanometers gives blue, 3 nanometers green, and 5 nanometers red. With the addition of red, not available in conventional phosphors, a cheap gallium nitride blue LED's color rendering index reaches 91 (near incandescent), looks warm to eyes, and overall efficiency remains high at over 60 lumens per watt. Apples look a rich red again, and LEDs start looking like a smart wager.

Bulovic and partners have formed a spinoff startup company named QD Vision, which brought quantum dots to the market in 2009. The technology is now being applied to LED displays, concentrating solar PV cells, and other military needs.

LEDs represent a sustained R&D program that has truly delivered: a no-regrets, enabling technology that has progressed from niche applications to the mainstream. Academic and industry efforts have together resulted in this generation's contribution to a long, unbroken arc of gains in what is perhaps humanity's defining technology: today's household LEDs are approximately 30 times as effective at producing light as the first electric filament lamps, 300 times as efficient as a kerosene lantern, and 1800 times as efficient as a Babylonian sesame oil lamp. Put another way—as estimated by the economist William Nordhaus—a typical laborer at the outset of the 19th century needed to work 40 hours to afford enough whale oil candles for the equivalent light output of an evening's use of a single incandescent bulb; an American citizen today, with an LED-equipped lamp fixture, earns that much illumination every second. But rather than meaning that the research is complete, successes to date in the manipulation of light are pushing researchers toward new boundaries of knowledge.

For example, while the energy efficiency of LEDs already greatly outperforms that of conventional illumination options, researchers are finding that we are still far below theoretical performance limits. While the internal quantum efficiencies of some LED designs can reach near 100 percent, the external efficiency—that is, useful light delivered—is often only a few percent. Realizing this headroom will be especially important as global access to electricity and the demand for lighting grows. Radically lower-power thin-film LEDs will also open the door to new, embedded uses in a variety of mobile, aerospace, medical, and military applications.

And just as LEDs have offered a step-change in today's lighting technology, equally compelling advances in the fundamentals of how we are able to control light itself are now just visible on the horizon. Namely, the ability to manipulate individual photons of light at the nanoscale to the same degree that our electronics industry now guides electrons opens up a huge new front for research and development on sensing and high-speed communications: replacing electricity with light.

ON THE HORIZON   Georgia Tech's high-efficiency piezoelectric microwire LEDs

Researchers have used zinc oxide microwires to significantly improve the efficiency at which gallium nitride LEDs convert electricity to ultraviolet light. The devices are believed to be the first LEDs that will have performance that has been enhanced by the creation of an electrical charge in a piezoelectric material using the piezo-phototronic effect.

Because of the polarization of ions in the crystals of piezoelectric materials such as zinc oxide, mechanically compressing or otherwise straining structures made from the materials creates a piezoelectric potential—an electrical charge. In the gallium nitride LEDs, Georgia Institute of Technology researchers used the local piezoelectric potential to tune the charge transport at the p-n junction—the diode formed where two semiconductor materials interface. The effect was to increase the rate at which electrons and holes recombined to generate photons, enhancing the external efficiency of the device through improved light emission and higher injection current.

The devices produced increased their emission intensity by a factor of seventeen and boosted injection current by a factor of four when compressive strain of 0.093 percent was applied to the zinc oxide wire. "By utilizing this effect, we can enhance the external efficiency of these devices by a factor of more than four times, up to eight percent," said Georgia Tech material science and engineering professor Zhong Lin Wang. "From a practical standpoint, this new effect could have many impacts for electro-optical processes—including improvements in the energy efficiency of lighting devices."

Credit: John Toon *for Georgia Tech Research News ("Boosting LED Efficiency: Zinc Oxide Microwires Improve performance of Light-Emitting Diodes (LEDs) through the Piezo-phototronic Effect," October 31, 2011)*

ON THE HORIZON   Stanford's low-power polariton laser

Lasers are an unseen backbone of modern society. The physics powering lasers, however, has remained relatively unchanged through fifty years of use. Now, an international research team including Stanford professor Yoshihisa Yamamoto and research associate Na Young Kim has demonstrated a revolutionary electrically driven polariton laser that could significantly improve the efficiency of lasers.

All lasers are based on Einstein's principle of stimulated emission. Charged particles, such as electrons, exist in discontinuous energy levels like rungs on a ladder. An electron provided with enough energy can become excited and "jump" up to a higher energy level. Excited electrons can spontaneously fall down to an available lower energy level, shooting off the difference in energy as a bit of light called a photon. The process, however, is inefficient: conventional lasers waste energy, unnecessarily exciting electrons to higher energy levels even when the lower levels are too full to accept the excited electrons when they fall.

A polariton laser, however, pairs electrons with so-called "holes" to form another type of particle, an exciton. These excitons are bosons, and an unlimited number of them can inhabit any given energy level. Using bosons in lasers has been a scientific goal for decades, but Yamamoto's team is the first to successfully build an electrically driven laser using bosons. This change drastically reduces the amount of power required to run the laser. The current iteration of the polariton laser requires two to five times less energy than a comparable conventional laser, but could require 100 times less energy in the future.

"We're hoping we can replace conventional semiconductor lasers with these polariton lasers in the future," Kim said. The device is already being utilized by Stanford researchers developing quantum computers and quantum simulators.

Credit: Thomas Sumner *for the Stanford News Service ("Stanford Physicists Develop Revolutionary Low-power Polariton Laser," May 20, 2013, in the Stanford Report)*

ON THE HORIZON　Stanford's synthetic magnetism photon control

An interdisciplinary team of physicists and engineers has created a device that tames the flow of photons with synthetic magnetism. In fashioning the device, the team has broken what is known in physics as the time-reversal symmetry of light. Breaking time-reversal symmetry in essence introduces a charge on the photons that reacts to the effective magnetic field the way an electron would to a real magnetic field.

For engineers, it means that a photon traveling forward will have different properties than when it is traveling backward, the researchers said, and this yields promising technical possibilities. "The breaking of time-reversal symmetry is crucial as it opens up novel ways to control light. We can, for instance, completely prevent light from traveling backward to eliminate reflection," said Stanford professor Shanhui Fan.

The Stanford solution capitalizes on recent research into photonic crystals—materials that can confine and release photons. To fashion their device, the team members created a grid of tiny cavities etched in silicon, forming the photonic crystal. By precisely applying electric current to the grid, they can control—or "harmonically tune"—the photonic crystal to synthesize magnetism and exert virtual force upon photons.

The researchers reported that they were able to alter the radius of a photon's trajectory by varying the electrical current applied to the photonic crystal and by manipulating the speed of the photons as they enter the system. Providing a great degree of precision control over the photons' path, this dual mechanism allows the researchers to steer the light wherever they like. In essence, once a photon enters the new device it cannot go back. This quality, the researchers believe, will be key to future applications of the technology, as it eliminates disorders such as signal loss common to fiber optics and other light-control mechanisms.

Credit: Andrew Myers *for the Stanford School of Engineering ("Taming Mavericks: Stanford Researchers Use Synthetic Magnetism to Control Light," November 1, 2012, in the Stanford Report)*

Rafael Reif on Game Changers
President of MIT

You could look around this room and see a gathering of the most creative and influential individuals in energy research today. Or you could look at this room and say the innovators here and technologies they represent are an incredible manifestation of the value of the American research university. As a group, our innovators would not offer the nation the same value—the capacity to tackle these tremendously complex and difficult projects—if they were all to work independently in their own garages. The scientific advances that emerge from their labs are born from the creative collaboration of extensive teams of faculty, researchers, and students. The work requires daring experimentation sustained over months and years. It depends on extraordinarily complex and costly equipment and it is worth every penny.

If the United States wants solutions to the grand challenge of a sustainable energy future, I believe it needs to invest more, and not less, in the research engine of its finest universities. Certainly, some of that research can and should come from industry and other sources. But much of it must come from universities, where researchers have the creative freedom of an academic life.

Today, as the world faces great problems around energy and climate, and related challenges around water and food, society needs the creative force of its universities more than ever. So let us celebrate the brilliant tradition of the federal-university partnership that brought us this far. And let us commit to doing everything in our power to preserve it so we can help the nation achieve a sustainable energy future.

KNOWN UNKNOWNS

Of course, not all energy research is focused on delivering "charismatic" consumer-facing products like electric cars or solar panels. Often, science and engineering advances have cross-cutting implications that are hard to pin down, but become fundamental building blocks of the advanced products we ultimately want. For example, an application of the same nanoscale surface property that can prevent a plane crash might also make a better ketchup bottle or improve the efficiency of a coal-fired power plant. New computing algorithms could make your smartphone's battery last a week, or they could be used to recover from a terrorist attack on the nation's electric grid. The biggest changes can come from places that few people expect, and long-shot R&D efforts may in fact one day have the biggest payoffs.

The exciting thing about broadly applicable research like this is that we may have the answers to tomorrow's game changers already in hand today. Union Pacific Resources' experiments with adding polymers to water to reduce friction and improve well flow was not hailed as a major breakthrough until Mitchell

Energy applied it to its shale gas acreage in the Barnett as "slick water" fracturing. In the same way, some of the "known unknowns" presented here, useful in their own ways already, may someday prove to be part of the biggest game changers of all.

> What are the game changers that are inherently "left field"? Have we not covered something that we don't see—or that is coming around the bend but it's not quite there yet—that will create discontinuities in our energy landscape and trajectory?
>
> —Arun Majumdar, *Google vice president for energy**

Thermoelectrics and Thermionics

Thermoelectrics and thermionics deal with better understanding and manipulating the production of electricity from heat in solid-state materials. This is applicable across the energy spectrum. For example, as an offshoot of conventional PVs, thermionic solar devices could be used to harvest both the heat and light energy of the sun, producing power at high efficiencies across various temperature ranges.

> We have work that would combine the solar cell with something like concentrating solar power. . . . And it's looking very, very promising. I think one of the great things about having a lot of people interested in this area is there are a lot of really good ideas coming out, resulting in all kinds of possibilities.
>
> —Nicholas Melosh, *Stanford associate professor of materials science and engineering*

Elsewhere, research into improved thermoelectric devices aims to increase the applicability of using solid-state materials to directly create electricity from waste heat, with or without light. The principle is well known—the Voyager spacecraft, for example, use thermionic power generators fed by the decay of radioactive isotopes rather than solar panels in the darkness beyond our solar system. Improving the reliability, cost, and useful temperature range of such devices could, however, make them more useful in improving supply-side energy efficiency in new places where it is impractical to apply conventional, large-scale, steam-based cogeneration technologies.

Ongoing crosscutting R&D on thermoelectrics and thermionics

• Scientists working at Stanford have improved an innovative solar-energy device to be about 100 times more efficient than its previous design in converting the sun's light and heat into electricity, and they expect to achieve at least another tenfold gain in the future. The device is based on the photon-enhanced thermionic emission (PETE) process first demonstrated in 2010, which uses a special semiconductor chip to make electricity by using the entire spectrum of sunlight, including wavelengths that generate heat. Because the efficiency of thermionic emission improves dramatically at high temperatures, adding PETE to utility-scale concentrating solar power plants may increase their electrical output by 50 percent.

Credit: Mike Ross *for the SLAC National Accelerator Laboratory ("A Solar Energy Chip 100x More Efficient," March 18, 2012)*

• Thermoelectric materials can be used to turn waste heat into electricity or to provide refrigeration without any liquid coolants, and a research team from the University of Michigan has found a way to nearly double the efficiency of a particular class of them made with organic semiconductors. Organic semiconductors are carbon-rich compounds that are relatively cheap, abundant, lightweight, and tough. But they have not traditionally been considered candidates for thermoelectric materials because they have been inefficient in carrying out the essential heat-to-electricity conversion process. U-M researchers improved upon the state-of-the-art in organic semiconductors by nearly 70 percent, achieving a figure-of-merit of 0.42 in a compound known as PEDOT:PSS.

Credit: Amy Mast *for the University of Michigan Energy Institute*

• Caltech scientists have recently concocted a recipe for a thermoelectric material that might be able to operate off nothing more than the heat of a car's exhaust. The thermoelectric semiconductor material they reported shows high efficiency at about 125 to 626 degrees Celsius—less extreme temperatures than commonly encountered with similar applications in spacecraft and power plants. By adding a small amount of selenium to a lead telluride compound, researchers created regions called "degenerate valleys" in the crystal's electronic structure. These were arranged in such a way as to provide a more favorable pathway for charge carriers to follow, a trail of equal-energy stepping stones through the material. In addition, adding the selenium creates multiple regions called point defects, which makes heat dissipate more slowly through the material. Ultimately, this balance of low electrical resistance and high thermal resistance improved the effectiveness of these new materials by roughly twice. Using this solid-state material to charge a car's electrical system in place of today's mechanical alternators could potentially improve a car's fuel efficiency by 10 percent.

Credit: Dave Zobel *for Caltech Media ("Caltech Researchers Develop High-Performance Bulk Thermoelectrics," May 23, 2011)*

• Research on thermoelectric devices out of the University of California at Berkeley and Lawrence Berkeley National Laboratory is being put to work in North American natural gas fields. The key innovation is a way to produce thermoelectric materials from relatively cheap and abundant silicon rather than conventionally used but rare compounds such as bismuth telluride. Until now, the relatively high thermal conductivity of silicon has precluded its use in thermoelectrics—the material does not allow for the many hundred-degree temperature gradients to form on opposite ends of the device that are necessary to induce an electric current. Pairing nanostructured materials such as silicon nanowires along with new electricity-gathering metal contact designs has helped overcome that. Development of the Berkeley technology is continuing under a startup company; an early commercial application targets passive harvesting of waste-heat into useful electricity at Canadian natural gas production wells.

Surfaces at the Nanoscale

New tools increasingly allow scientists from a variety of fields to observe, manipulate, fabricate, and model the behavior of materials at the nanoscale. At the same time, a growing body of experimental results and computational models is helping to make nanoscale dynamics more accessible. One result of this has been the development of new fields of inquiry into the importance of material surface textures at the nanometer scale.

For example, there are a lot of places—in energy and elsewhere—where making things more slippery makes them work better. Researchers would like to avoid methane hydrate crystals from clogging flow in cold, undersea pipelines. Similarly, keeping winter ice off power lines helps avoid debilitating blackouts. Airplane wings can ice over, which is both dangerous and contributes to snow delays on the tarmac. The operation of desalinization plants is limited by their ability to effectively shed the fresh water they produce. And for the steam turbines that drive generators to produce the vast majority of the planet's electricity, a more water-repellent surface means higher efficiency and more durability.

A related research avenue seeks to better understand how nanoscale and microscale roughness affects a material's ability to transfer heat. This is important, for example, in power plants that typically use phase change—for example, boiling water—to dissipate heat, or for the computer chips, where power densities are already ten times that of a stove top.

Nanoscale surfaces also play a role in material lifetimes—for example, resistance to heat, corrosion, or mechanical stresses, and even the impact of radiation damage. This relates directly to the sustainability of the United States' extensive and low-polluting light water reactor commercial nuclear power fleet. Understanding material degradation is part of further extending plant lifetimes from previously forty, to now sixty, and in the future eighty years while safely increasing both rated output and yearly uptime. Already, nuclear fleet operational improvements have increased the average capacity factor of the US nuclear fleet to over 86 percent in 2012, up from 66 percent in 1990. This improvement is akin to having built thirty new nuclear plants during a period when essentially no new steel was actually put into the ground.

Light water reactors will be the workhorse of nuclear energy for at least the remainder of this century and therefore should get more attention. . . . How do you make the most out of existing nuclear power plants? We need to help them age gracefully and determine whether there are important things that have to be done now before their licenses are extended beyond 60 years. An example of a very wise enhancement is to deploy more accident tolerant fuel.

We have now accumulated substantial evidence that the radiation doses to the vessels are much lower than doses that would cause their embrittlement. Therefore, the extension of life from forty to sixty years has in fact started already. What people have in mind is going from sixty to eighty years, but then they ask the question not whether the vessel material can take it but whether the concrete that surrounds the vessels can take it, and whether the instrumentation and control equipment should be updated.

—Mujid Kazimi, *MIT TEPCO professor of nuclear engineering and director of the Center for Advanced Nuclear Engineering Systems*

Ongoing crosscutting R&D on surfaces at the nanoscale

- New MIT research is developing water-repellent coatings to improve the efficiency and durability of steam turbines. The so-called "LiquiGlide" coating started with the conventional approach to making a surface non-wetting: give it nanoscale roughness and then coat it with a low-surface-energy material, usually a polymer. Air pockets are then trapped between the bumps, serving to reduce the contact of the liquid with the surface so it flows off more easily. It then went one step further: to prevent these air pockets from filling with water, the MIT Varanasi Research Group filled the spaces between the surface textures not with air, but with a liquid lubricant that will not mix with the material being repelled. The resulting structure is durable, resistant to aerodynamic assault, and unfazed in a vacuum. The new coatings can be applied to materials ranging from glass and plastics to metals and ceramics, and by adding the lubricant to a textured surface, the researchers can get condensed water droplets to move 10,000 times faster than without it.

Credit: Nancy Stauffer, ©*Massachusetts Institute of Technology, used with permission (MIT Energy Initiative "Novel Slippery Surfaces: Improving Steam Turbines and Ketchup Bottles," June 20, 2013)*

- Researchers at MIT have found that relatively simple, microscale roughening of a surface can dramatically enhance its transfer of heat. Such an approach could be far less complex and more durable than approaches that enhance heat transfer through smaller patterning in the nanometer (billionths of a meter) range. The team concluded that the reason surface roughness greatly enhances heat transfer—more than doubling the maximum heat dissipation—is that it enhances capillary action at the surface, helping keep a line of vapor bubbles "pinned" to the heat transfer surface. This delays the formation of a vapor layer that would greatly reduce cooling. The new research also provides a theoretical framework for analyzing the behavior of such systems, pointing the way to even greater improvements. A major potential application is in computer server farms, where the need to keep many processors cool contributes significantly to energy costs.

Credit: David Chandler, ©*Massachusetts Institute of Technology, used with permission (MIT News Office "Better Surfaces Could Help Dissipate Heat," June 26, 2012)*

• Experimental and computational research at MIT are helping to characterize stress corrosion cracking of materials within the extreme environments of nuclear power plants. Through a process called nanoindentation, using the customized tip of a scanning electron microscope, researchers have been able to observe how dislocations associated with crystalline grain boundaries increase chemical reactivity and physical corrosion. They are also developing new computational tools that can shed additional light on the corrosion process as well as on radiation damage. This work focuses on zirconium alloys used as nuclear fuel cladding, an outer tube that surrounds the ceramic fuel pellets inside today's reactors. Since thinning of the cladding raises safety concerns, power plant operators use computer models to predict how fast damage from corrosion will progress. Models incorporating these new computational methods should help operators more accurately predict the impact of corrosion and radiation damage on the aging and performance of the fuel used in today's power plants.

Credit: Nancy Stauffer, ©*Massachusetts Institute of Technology, used with permission (MIT Energy Initiative "Stress Corrosion Cracking," December 19, 2012)*

Algorithms and Software

I believe there's a game changer, which is in algorithms. Software has to be smarter somehow, and do a better job than it does right now. There's a question of whether we're ever going to get there—whether we're going to be repeatedly attempting to modernize the grid, like we are with the air traffic system—or whether we're going to magically make an Internet-like system, which just works.

—Sanjay Lall, *Stanford associate professor of electrical engineering and of aeronautics and astronautics*

Algorithms are the step-by-step, repeated data-handling procedures that are the building blocks of computer software. Algorithms are the "secret sauce" that make Google's web search so good and let your handheld calculator solve an equation. Algorithms encrypt digital documents and they are used to sort through massive amounts of data to help arrive at a decision. Algorithms are everywhere that computers are, and developing new ones can help do everything from making computers do what we want using less energy and heat, to letting us better rely on computers to make life- and money-saving decisions in faster-than-human scale environments such as our electric grid.

For example, 2 percent of all US electricity is used in remote data centers. That figure is set to grow dramatically as we move to cloud computing and increase overall computational needs. More energy-efficient computation could not only reduce energy use, but also make our consumer electronic devices more useful: smaller batteries, longer lifetimes, and faster processors less bound by thermal constraints. Just as Moore's Law describes the trend of increasing computing power over time, computing

energy efficiency has also increased steadily for decades: compu-
tations per kilowatt-hour double roughly every eighteen months.
At the same time, as chips gets physically smaller, the power
density of microprocessors has approached that of a nuclear
reactor. One approach toward keeping up efficiency trends is to
rethink computing algorithms themselves from the ground up
to overcome concerns that progress on the physical chip-side
could saturate.

> We are wildly increasing the exposure surface of our infrastructure,
> through increasing the amount of the control systems that we
> have, and things like smart meters, network buses, and other
> interfaces. And you don't have to be very adventurous or very
> creative to imagine ways that this could go horribly wrong. It's worth
> thinking about the cyber security of the complexity we're getting
> ourselves into.
>
> —Julio Friedmann, *Lawrence Livermore National Laboratory chief
> energy technologist**

At the opposite end of the spectrum, the increasing IT con-
nectedness of industrial control systems—power plants, pumping facilities, pipelines, electricity transmission—on top of a
"smarter," information-dense grid demands that we put new
control algorithms into service. The independent system operators that manage the flow of electricity across the grid already
rely on computer software to make centralized real-time generator dispatch and routing decisions. Intelligently defend-
ing our energy infrastructure from faults or malicious attack
will require handing over system monitoring to a new, robust,
automated—and likely decentralized—operating system that
can respond faster than any human can.

The electric grid is a complex, distributed, stochastic, and nonlinear dynamical system. It is the nightmare of control theorists, and I'm a control theorist.

—Munther Dahleh, *MIT Engineering Systems Division director and professor of electrical engineering and computer science*

Ongoing crosscutting R&D on algorithms and software

- MIT research is aiming to ensure future improvements in the energy efficiency of computing by radically rethinking algorithms—the step-by-step procedures at the core of all software programs. Various algorithms may solve a given computational problem, but they may differ substantially in the amount of time, memory, and energy they use to do it. The researchers' approach is based on "reversible computing," an idea first proposed in the 1970s. Using specially devised theoretical models, they have spent months analyzing basic algorithms to see whether they can be made reversible—or more reversible. Early results point toward a huge potential for energy savings in procedures used for processing big data, such as when running network routers or performing web searches. The ultimate goal of this complete rethinking of algorithms is to have computers spend a million times less energy per computation.

Credit: Nancy Stauffer, *©Massachusetts Institute of Technology, used with permission (MIT Energy Initiative "Energy-efficient Computing," June 20, 2013)*

- The University of Southern California is working with the Los Angeles Department of Water and Power to deploy its "DETER Testbed" to demonstrate next-generation cyber security for the electric grid. DETER (Defense Technology Experimental Research) is a collaborative experimental platform used by hundreds of researchers worldwide to emulate a variety of Internet-based attacks and other uses of malicious code. The system allows for repeatable, measurable experiments that can be used to evaluate and improve network architecture security for connected infrastructure. Current research aims to evaluate the security of electric utility customer information, the risk of two-way IP-based communication to US electricity infrastructure, and the ability of the electric grid to defend itself from new cyber threats.

Catalysts and Computational Chemistry

There's a lot of work left. There's a lot of biology that has to be understood and manipulated. But nevertheless, it's a start. Some of them are easy things, some of them are really hard things. But that's where we've got to start.

—Arun Majumdar, *Google vice president for energy**

Synthetic fuels, biofuels, flue gas carbon capture. Each category on its own is a potential game changer. And success in each demands a new command of chemistry. One subarea in particular stands out: the identification and use of exotic, potentially never-before formulated catalysts to drive chemical reactions faster and at lower activation energy levels. Catalysts are central to any number of chemical reactions in any field, but the search for new and effective catalysts in the energy sector has taken on a new urgency as we increasingly seek to cheaply and reliably convert organic compounds, such as hydrocarbons, and to do so at ambient conditions.

As just one example, catalytic reduction of atmospheric CO_2 to methane and higher hydrocarbons is a basic research thrust with a potential for major long-term payoffs—pulling fuel from the atmosphere and scrubbing CO_2 to boot. Conventional fuel cell chemistry depends on expensive platinum catalysts to oxidize hydrogen fuel and produce electricity, so we would like to identify and test new catalysts that can do the same job more cheaply. This would mean, for example, moving from today's 12 percent efficient platinum hydrogen production systems at $40,000 per square meter to earth-abundant material-based systems at $100–$200 per square meter. Plant photosynthesis itself is about a 1 percent efficient process, and experimental earth-abundant systems today are about 2 percent efficient, but there remains much work to make such technology a reality. New synthetic catalysts could also help improve the efficiency of solvent-based CO_2 flue gas capture systems or the conversion of the United States' newly abundant natural gas into a more convenient liquid hydrocarbon fuel. The applications are extremely broad.

> Whoever can develop a way of taking methane and coupling it to ethylene is going to have a huge advantage in exploiting natural gas. And that's what we set out to do in the last couple of years. . . . We test 100 catalysts a day for their ability to do oxidative coupling of methane. . . . And in a year, we improved performance by about 5x and have several different catalysts that are economically viable for oxidative coupling of methane.
>
> —Angela Belcher, *MIT W. M. Keck professor of energy*

The urgent need for new catalysts and other compounds in the energy sector has already led to a revolution in the scientific method used to identify and evaluate them. Traditionally, researchers would take a number of chemical compounds and then test them in the lab, one-by-one, for the desired reactive properties. Today's high-throughput computational modeling of various formations and molecular permutations reverses that process, allowing the scientist to first screen various options for desired functionality, before establishing model electric forms, and only as the last step synthesizing the chemical composition itself. This approach, combined with shared cloud-based chemical datasets, now allows a single gas separation lab to go from experimenting on just dozens to instead of millions of catalysts—for example, metal organic frameworks—per year.

> The very term "game changer" applies to this category of solar fuels. It is clearly in the long-term category, because this is going to take not one but several major breakthroughs to be feasible. But the idea that taking photons, either as solar-derived electricity or via direct photocatalysis to convert water to hydrogen to hydrocarbon or reduce CO_2 to useful organics, is truly an energy game changer.
> —Robert Armstrong, *MIT Energy Initiative director and Chevron professor of chemical engineering*

Many scientists now note that early private-sector funding of university basic science R&D efforts in these areas over the past decade, at a time when the federal government was not yet deeply involved, helped to create an entire research infrastructure—university programs, energy research centers, student interest, and students graduating into industry. It allowed those who were already deeply committed to perform their own research with fewer struggles to obtain funding, and it also drew in others who might not have otherwise been involved at all. The end result is that there is now a huge capacity to scale R&D in these on-the-horizon—but potentially large payoff—areas that was not there a decade ago.

Ongoing crosscutting R&D on catalysts and computational chemistry

- Scientists from SLAC and Stanford have teamed with researchers in Germany's Fritz Haber Institute to discover a key part of the most common process for making methanol. Their experimental evidence reflects the full complexity of the reactive active sites of the sponge-like copper, zinc oxide, and aluminum oxide industrial catalysts that are highly efficient at making methanol from syngas today. Namely, their work revealed that under real-world conditions, the catalyst's copper surface was folded into "steps" and decorated with particles of zinc oxide, and that this configuration was stabilized by other defects in the material. With this information in hand, the researchers are in a position to further tweak the recipe and perhaps find an efficient way to make methanol from the CO_2 that is produced by burning fossil fuels.

Credit: Glennda Chui *for the SLAC National Accelerator Laboratory ("Study Cracks a Secret of Methanol Production," May 23, 2012)*

- MIT chemical engineers have devised a cheaper way to synthesize the key biofuel component, gamma-valerolactone (GVL), which could make its industrial production much more cost effective. GVL is attractive because of its versatility: it has more energy than ethanol, it could be used on its own or as an additive to other fuels, and it could also be useful as a "green" solvent or a building block for creating renewable polymers. To get around cost-prohibitive aspects of the conventional manufacturing process that rely on precious metals, the team developed a new series of cascading reactions slightly different from the traditional pathway. The catalyst for this series of reactions is a zeolite—a porous silicate mineral containing zirconium and aluminum, both abundant metals. The entire process takes place at a relatively low temperature of just 120 degrees Celsius, and it does not require hydrogen gas. This makes the capital costs for the necessary equipment lower than they would be using the traditional process.

Credit: Anne Trafton, *©Massachusetts Institute of Technology, used with permission (MIT News Office "Making Alternative Fuels Cheaper," June 16, 2013)*

• A multidisciplinary team of scientists and engineers at the Energy Biosciences Institute at the University of California Berkeley has developed a new "hybrid process" for producing diesel and jet fuels from sugars by using a combination of biological and chemical transformations. The team identified catalysts and reaction conditions that lead to the highly efficient conversion of a mixture of acetone, butanol, and ethanol (ABE) to a family of hydrocarbons that can be used as diesel and jet fuels. The ABE mixture was produced by fermenting sugars using improvements to a process that had been used commercially for about eighty years to produce the individual components, but which had been discontinued. Thus, there do not appear to be major hurdles to the bioconversion aspects of the overall process. Although several small companies are trying to commercialize other routes from sugar to diesel or jet, the ABE-derived fuels appear to have the highest conservation of the energy inherent in sugar feedstocks.

Credit: Chris Somerville *for the UC Berkeley Energy Biosciences Institute*

THE US MILITARY

The Defense Department has funded an astonishing amount of today's most remarkable technology going back decades.
—Susan Hockfield, *MIT president emerita*

Many of the technologies and developmental efforts discussed in preceding chapters also have significant military and national security implications. The ongoing efforts of the military to dramatically reduce its energy consumption are both driving sponsorship of energy R&D and simultaneously allowing the DoD to act as an "early adopter"—piloting the use of civilian energy innovations prior to more widespread commercial use. Combined, Defense-wide spending in 2012 on energy research, development, testing, and demonstration alone was $1 billion—though even this was but 1 percent of its total innovation budget. Of course, there is strong precedent here. The US military—from R&D, to procurement, and a massive ability to scale and deploy technology—has long been an early supporter of game-changing technological innovations, including internet telecommunications, global positioning systems, and solid-state transistors.

I could not be more proud of what the military has done over these past ten years. And a lot of it, tragically, has been driven by the realities of combat. Energy, the Defense Department being a huge user of it, is what you live and fight and operate on. But it was clear early on, in the beginning of the war in Iraq, that energy was also costing us lives, because for every fuel convoy, at least one young American risked giving his life.

—Admiral Gary Roughead (Ret.), *Hoover Institution Annenberg distinguished visiting fellow and former chief of naval operations**

The federal government stands as the largest end user of energy in the United States with the DoD as the top consumer and accountable for a substantial percentage of that use. For example, US Navy petroleum consumption (for ships, aircraft, tactical vehicles, and generators) represents 0.4 percent of the use of petroleum across the United States. Dramatically reducing US military energy use—on base and in the field—can significantly enhance the security of the armed forces and the nation.

Any president is going to want a military that's ready, right now, for a global mission—that can deploy anywhere in the world rapidly for a big range . . . of missions, whether it's humanitarian and disaster relief, which the Department now considers to be core missions, or whether it's conventional combat, or irregular combat, or cyber war. We need to be ready for a full range of contingencies everywhere around the world. And that inherently requires a great deal of energy.

[W]e do need to find ways to use less energy for everything that we need, to get more military output for every unit of energy input.

—Sharon Burke, *Assistant secretary of defense for operational energy plans and programs*†

Extreme energy reliability and performance demands put the US military at the leading edge of implementing energy innovation. Where there is overlap with civilian needs, technologies that may still be "near at hand" for other commercial markets may already be deployed and useful on military bases and in forward-deployed locations. Other military energy innovation may be more narrowly targeted at a particular tactical edge—for example, directed energy weapons or rail gun technology. Taken together, the DoD, along with the DOE and other interagency partners, stands well positioned to lead, develop, and implement emerging energy strategy initiatives. Just as the US Navy's interagency, industry, and national leadership drove the development of nuclear energy in the twentieth century, the DoD's unique position as a sea, air, and land consumer of energy offers a rich opportunity to operationalize viable alternate fuels, cleaner energy sources, or game-changing energy technology.

Our primary platform is the soldier, the dismounted soldier. A lot of the fuel is what they're carrying on their back. And a lot of that is batteries. So that's where our focus is. We know we need to increase energy efficiency, so that is certainly a huge focus for us in many ways.

The dismounted soldier carries as many as seventy batteries of a dozen different types. We are in need of rechargeable batteries and energy storage for a variety of applications.

—Katherine Hammack, *Assistant secretary of the Army for installations, energy and environment**†

AVAILABLE TODAY

Navy's stern flaps

The Navy began installing stern flaps in 2009 on amphibious ships and other combatants in an effort to make ships more fuel efficient and save up to $450,000 in fuel costs per ship annually. Stern flaps are passive surfaces, analogous to the spoiler on a race car or airplane winglets, that induce planar flow around the ship's hull, reducing the drag and visible signature otherwise created from wake vortices. They are an excellent example of the Navy incorporating a proven, commercially available fuel-saving technology.

Navy's hull coatings

Marine growth such as biofilm or barnacles adds weight and increases drag. The Navy estimates that subsequent reductions in ship fuel efficiency and increased maintenance needs cost $1 billion annually, reduce vessel speeds by up to 10 percent, and potentially compromise acoustic stealth. Hull biofouling-prevention coatings therefore reduce costs and emissions while improving operational capability. And while many conventional biofouling treatments rely on toxic biocides, ongoing R&D aims to achieve more effective and environmentally benign results. Examples include biomimetic hull surface patterning and a class of dipolar ionic molecules that naturally exhibit both positive and negative charge. Advances here are particularly compelling as they could also benefit the broader global commercial shipping industry.

Navy's hybrid-electric drives for large combattants

The Navy has incorporated fuel-efficient hybrid-electric propulsion technology onto several of its next-generation big-deck amphibious assault ships: for example, the USS America (LHA-6) and the USS Tripoli (LHA-7), which are part of what the Navy calls its now-in-development, America-class amphibious assault ships. The hybrid drive allows the ship to propel itself using either electric motors paired to a diesel generator or a traditional gas turbine engine. Doing so reduces fuel use—thereby extending operational time between refuelings—by allowing the gas turbine to spend more time operating in the higher power bands where it is most efficient: electric motors help propel the ship at speeds up to around 12 knots, while the conventional gas turbine engines take over at higher speeds. At the same time, the diesel generators that feed the hybrid drive's electric motors can more efficiently provide onboard standby power for many of the ship's systems such as sensors, weapons, and other electronics. In 2009, USS Makin Island (LHD-8) became the first US Navy amphibious assault ship to feature a unique hybrid propulsion system that relies on two large gas turbines or two diesel electric motors. Arleigh Burke (DDG-51)-class destroyers are also now set be retrofit with hybrid-electric propulsion technology.

Navy's voyage planning software
Smart Voyage Planning (SVP) is a capability deployed as a software application: it uses fuel curves, weather, and ocean current data to plan optimal transit routes that minimize fuel usage. SVP capitalizes on real-time data and computing power to plot routes that have the potential to save 4 percent in annual fleet fuel cost.

US Marine Corps' ExFOB
The US Marine Corps' Experimental Forward Operating Base (ExFOB) collection of equipment is focused on water and energy efficiency when establishing a FOB during expeditionary operations at the small-unit level. Energy conservation systems range from lightweight solar panels to innovative adapters, mitigating the need to carry several batteries. These systems help to lessen the Marine Corps' dependence on liquid-fuel generators and logistics requirements.

Army's soldier power management systems
The Army has deployed the Squad Power Manager, a charging system that provides a centralized power source and access to the variety of man-packable equipment that soldiers now carry: GPS, multiple radio systems, night vision, and PDAs. Each can be connected to standard-issue wearable batteries alongside lightweight 10- and 20-watt solar blankets that weigh just a few ounces. This system improves flexibility and reduces the need to carry multiple, semidepleted batteries for each device.

The US military presents both unique challenges and opportunities in the energy domain. No other organization operates at such large scale—with industrial, commercial and residential facilities, sea and land bases, ships, submersibles, submarines, aircraft, and large vehicle fleets in every region, and in every environment around the world. And all of these operate within varying environmental, energy, and fuel constraints. At the same time, no other organization possesses the breadth and depth of experience in dynamic energy use and multisource energy production—including nuclear, geothermal, fuel cell, and advanced storage technology. Arguably, the DoD offers the world's best organizational, cultural, and intellectual platform for analysis and operational implementation of game changing technology and leading-edge infrastructure.

> A resource-efficient Marine is a more combat-effective Marine.
>
> —Colonel Robert Charette, *USMC Expeditionary Energy Office director**†

Recent Defense strategies explicitly acknowledge this nascent capability. For example, in 2009, Secretary of the Navy Ray Mabus announced for the first time five energy targets for the service. These goals exceeded US government mandates regarding alternative energy, renewable energy, and greenhouse gases and highlighted the US Navy's commitment to the imperative that alternative energy implementation is vital to national security.

One way to achieve these goals is through aggressive procurement of new or emerging energy technologies, which now totals half a billion dollars annually department-wide. For example, Naval Air Weapons Station China Lake illustrates how existing shore-based facilities are capturing the Navy's "conservation, efficiency, and alternatives" energy strategy. Here, the Department of the Navy oversees the operation of a

270-megawatt geothermal power plant that provides approximately 1.4 million megawatt-hours of electricity every year to California's electric grid. This facility is the largest renewable energy producer across the armed forces and is one of the largest geothermal electricity producers in the United States. The revenue from royalties of energy sales from this plant to California utilities goes directly to funding other renewable energy programs for the US Navy. This facility also now presents an excellent opportunity for the Navy to become a serious "early adopter" of energy security-enhancing renewable distributed generation coupled with next-generation microgrid infrastructure.

> The Air Force fights from fixed locations, our installations. . . . We have a significant dependence on the commercial grid for the energy that we use day-to-day on our installations. We all know that introduces a certain amount of risk, and certainly jeopardizes our effectiveness to do our critical missions even if a power outage is only for a couple of hours. So for us, we think in terms of installation security as one of our primary concerns.
>
> —Terry Yonkers, *US Air Force assistant secretary for installations, environment and logistics**†

Elsewhere, the US Navy has announced a $200 million investment in solar technology for bases in the southwestern United States. Implementing localized generation technologies—PV arrays, wind turbines, micro-turbines, reciprocating engines, fuel cells, combustion turbines, steam turbines, and combined heat and power—allows the Navy to bypass the centralized system of generation and dispatch and to meet its own electricity needs. At the same time, it helps stabilize and support the grid, mitigating vulnerabilities.

We have become utterly dependent on the electric grid over the last fifty to seventy years. Given our dependence on the grid, if it were to suddenly go away, it could have catastrophic results to our way of life. The problem is that it's becoming increasingly fragile and extremely vulnerable.

—Richard Andres, *National Defense University professor of national security strategy**†

In total, US Navy shore facilities today produce 12 percent of their total annual energy needs from renewable sources. Localized renewable energy sources and other distributed generation technologies connected to next-generation infrastructure are helping to build a foundation toward service energy goals, including deriving 50 percent of total energy consumption from alternative sources by 2020. On base and in forward deployment, the Navy is joining with the Marine Corps, Army, and Air Force in moving energy technology from the lab to the field.

NEAR AT HAND

Army's Consolidated Utility Base Energy system

Consolidated Utility Base Energy (CUBE) is an integrated power electronic platform for a 60-kilowatt PV-battery-diesel hybrid power system developed to provide power to forward operating bases. The modular CUBE prototype is intended to integrate four 15-kilowatt PV arrays, one 30-kilwatt battery bank, and two 30-kilowatt diesel generator sets to power a 60-kilowatt load. Onboard power electronics include PV Maximum Power Point Tracking (MPPT) converters, battery charge/discharge converters, and a three-phase inverter capable of smoothly transitioning between operation as a stand-alone voltage source and operation in parallel with the diesel generators or a utility grid connection.

Army's polymer conformal battery and SWIPES

The Army's Natick Soldier Research Development and Engineering Center is working to develop a 0.8-inch thick battery that can be placed directly into a soldier's vest with minimal added bulk. Similar to the Squad Power Manager, described above, such a battery could be integrated into the Soldier Wearable Integrated Power System (SWIPES), whereby a single battery powers a variety of worn devices through a network of internal cable routing and pockets. The goal of these systems is to help extend mission length while passively ensuring that each needed device remains charged and available for use. SWIPES has been named as one of the US Army's top ten innovations; field testing has begun on several hundred units through the Army Rapid Equipping Force and Project Manager Soldier Warrior.

US Marine Corps' concentrated solar harvesting technology

These concentrating solar harvesting systems produce power and hot water at geographically remote forward-operating bases, with the aim of reducing the area required to deploy solar systems with capacities of 5 kilowatts and below. In addition to more conventional solar water heating, the program also includes lens-focused PVs and solar thermal dishes. Water heating today generally relies on electricity produced from potentially wet-stacked diesel generators, likely operating below their efficient load levels, so incorporating solar power here can reduce the need for fuel delivery and maintenance.

US Marine Corps' tactical vehicle fuel efficiency

Over the past decade, while a typical Marine battalion's lethality has gone up, so has its energy use: 250 percent more radios, 300 percent increase in IT, 200 percent more vehicles, 75 percent increase in vehicle weight, and 25 percent decline in vehicle fuel efficiency. The addition of significant weight from armor and other warfighter requirements and continually increasing onboard power requirements for new electronic systems such as improvised explosive device (IED) jammers, radios, vision devices, and communication equipment results in greater fuel demand at FOBs. This represents a major cost, logistics, and safety issue that the Marine Corps aims to reduce.

For example, idling the Medium Tactical Vehicle Replacement (MTVR) can maintain 2.4 kilowatts of power to off-board equipment, but consumes an average of 0.8 gallon of fuel per hour. This is in comparison with a typical 10-kilowatt tactical generator, which consumes less than one gallon per hour. To help address this, numerous efforts now focus on exporting vehicle power at idle or static conditions.

What is the future for expeditionary warfare technologies? The solar backpacks, for example. Even if we pull out of Afghanistan shortly, we're going to continue developing our technology. But how do we proceed with these technologies when they're not being shipped over to theater within six months? They're going to be important for the future of the Marine Corps, the Army, the Air Force, and the Navy. We don't want to slow our progress. We have to anticipate our warfighters' needs and be ready, not be scrambling after the fact.

—Jackalyne Pfannenstiel, *Former US Navy assistant secretary for energy, installations and environment**†

While a decade of combat needs has driven many of the operational energy innovations we now see in the field, and a forward-looking procurement strategy is helping to deploy leading-edge energy security technologies more broadly, it is also now evident that game-changing energy innovations will play a central role in defining the nature of advanced defense operational capabilities over the long term. While these technologies remain early stage and may fail to come to fruition, their implications are already compelling. High-energy weapons, for example, may have science fiction origins but have also now successfully been field-tested shipboard off the coast of California.

Soup-to-nuts, long-term support for high-risk, high-reward R&D projects such as this is truly a domain in which the DoD is organizationally without parallel. Whether its innovations ultimately have civilian application or not, few other global institutions are in the position to deliver this sort civilization-scale change. Increasing attention on the world of energy, which is so central to broader social endeavors, is therefore something worth applauding.

ON THE HORIZON

Navy's seawater-to-fuel systems

The Naval Research Laboratory, the corporate R&D facility for the Navy and Marine Corps, is developing technology that could one day produce synthetic jet or bunker fuel while at sea from the surrounding water. Rather than using conventional electrolysis, the process recovers CO_2 and produces hydrogen from seawater using an electrochemical acidification cell (the CO_2 being largely bound in bicarbonates in seawater). The gases are then reduced and hydrogenated into an olefin through an iron-based catalyst and further refined into usable synthetic hydrocarbon fuels. The process has achieved CO_2 conversion levels of up to 60 percent and is undergoing testing in the Gulf of Mexico. Though energy input for the process is still quite high, the ultimate goal is to be able to improve at-sea fueling operations, diversify the operational supply chain, and enhance self-sufficiency while at sea.

Navy's high-energy pulse power requirements

High-energy pulse power requirements are driving research and development of new high-power electronics, leading-edge generation, and power-distribution technologies.

The High-Energy Laser (HEL) program aims to offer naval platforms enhanced, economical defense capability against air and surface threats—including swarms of small boats—and future anti-ship cruise missiles. This solid-state laser system, which the Congressional Research Service has described as a technological "game changer" in Navy tactics, ship design, and procurement, has been successfully tested on small drone targets through the marine layer at sea. It is currently being test-deployed at a forward-operating location in the Persian Gulf. Miniaturizing requisite power delivery systems remains a major R&D focus.

Meanwhile, the Counter-Directed Energy Weapons (CDEW) Program is exploring how to adapt to and defend against a similar set of hostile, directed energy marine weapons, including lasers and microwaves. This cross-disciplinary research program spans the domains of material science, optics, and high-energy physics. The Office of Naval Research, together with the Naval Postgraduate School, the US Naval Academy, the Naval Research Laboratory, and naval air, space, and surface warfare centers are similarly investigating basic research topics related to countering the threats that come from directed energy weapons systems.

In parallel, the Office of Naval Research is continuing to develop nascent electromagnetic rail gun technology that could someday replace many shipboard chemical propellant-fired projectiles. This disruptive, high-energy weapon uses an electromagnetic field to accelerate an otherwise inert metal projectile to over 5,000 miles per hour, turning it into a kinetic energy warhead. Navy rail gun research efforts are now entering their second phase, which focuses on improving repeated fire capability and supporting electronics for high-impulse power delivery. A 32-megajoule proof-of-concept device has been successfully demonstrated—an energy level that would be capable of delivering a 100-nautical mile projectile range.

Air Force's directed energy systems

The US Air Force Research Lab is also developing novel directed energy systems. Key research areas include laser systems, high-power electromagnetics, weapons modeling and simulation, and so-called "directed energy and electro-optics for space superiority." This research program draws upon experience gained through the previous development and testing of megawatt-class airborne lasers and the operation of ground-based large-diameter telescopes equipped with adaptive optics for space imaging. Related work includes the development of counter-electronics technologies that can precisely degrade, damage, or destroy hostile electronic systems.

ENERGY RESEARCHERS AS A STRATEGIC ASSET

The heart of our business, first, is research and development.
—John Deutch, *MIT professor of the institute*

The message from a careful look at the United States' university energy R&D work is clear. A lot has been done already that has been commercialized by the private sector. A lot more is sitting on the cusp—things close to commercialization that can make a big difference—and it is quite clear that progress will be made if we keep at it. And we know that there are other things that have a way to go—scientists and researchers themselves will be the first to say that—but they are nevertheless making conceptual progress, and getting this right is essential.

When I first arrived at MIT, I started talking to people about what our opportunities and responsibilities were for the next decade. Much to my astonishment—having been in academic institutions my entire life, I was expecting to get about 5,000 views from 1,000 faculty—there was almost complete unanimity that the most important thing for MIT to do was to make a significant difference in changing the world's energy system.

—Susan Hockfield, *MIT president emerita*

One thing that really stands out is how much universities have transformed their research capabilities in energy science and engineering over the past decade. In 2002, for example, Stanford spent $2 million per year on energy R&D; the government of Saudi Arabia funded more campus research than the US government did. Ten years later, the figure is over $60 million—supported by a broad swath of public sector, industry sponsors, and private donors. At MIT, MITEI was not even launched until 2006. In just the few years since, more than 800 research projects have been launched. In 2016 and 2026, we are going to begin seeing, in a dramatic way, the commercial fruits of this effort launched years before.

> As we've thought about ramping up the energy activities at Stanford, our goal from the beginning was always to get people from across the university focused on the problem. And I think what's been equally exciting to see is the dramatic growth in graduate student interest. . . . We've really brought together all different parts of the university to work on it. . . . At the same time, we've tried to develop collaborations around the country with a number of our colleagues. I think no one university can solve this problem.
>
> —John Hennessy, *Stanford University president*

The impact of the university transformation extends to the individual level as well. A quarter of MIT's faculty is engaged in energy research or education across all five schools and 22 of 25 departments. Stanford now counts more than 200 faculty members and an equal number of graduate students whose work touches on the subject of energy. Many of them are so focused on the energy problem as to have tied their academic careers to successes here. One well-known nanotechnology materials scientist, now a leader of the field, reports only turning to what became his energy subspecialty as a postdoctoral scholar after seeing the social challenges and research opportunities that were opening within the energy space. This story and the numbers behind it are no doubt repeated all over the country: over 300 faculty doing energy research at University of Texas at Austin, 130 faculty at the University of Michigan, 100 at Georgia Tech, 100 at Berkeley, 80 at the University of Southern California. And the list goes on.

> We're on the right track with the very significant energy research effort we have today in our research universities, but the key is to continue to innovate—to continue to think in terms of game changing technologies—because without this forward thinking, we will stagnate and fall short of the challenging goals ahead.
>
> —Robert Armstrong, *MIT Energy Initiative director and Chevron professor of chemical engineering*

These researchers, and the program infrastructure that supports them, are a national asset. They are an alchemic engine that takes in hard problems and moderate funding dollars at one end and at the other end produces the innovations that literally fuel our nation's progress. In an era in which so many resource issues have been framed in terms of scarcity, our energy researchers are discovering abundance. And while these academics may spend their days in labs, they do not work in a vacuum. Those who have chosen to undertake new explorations in energy over the past decade could leave for other bounties in the next. Perhaps more distinctly, it is breathtaking the number of field-leading innovators—and their thousands of graduate students—who have chosen to emigrate to US universities for their most productive years. This is a de facto vote of confidence in the US innovation system, but it cannot be taken as granted.

> I think that one of the great things the United States has are these fantastic research universities. They're the envy of the world and it's interesting how at many of them, a lot of emphasis has swung toward energy. That's a very good thing for the country.
>
> —George Shultz, *Chairman of the Shultz-Stephenson Task Force on Energy Policy and Thomas W. and Susan B. Ford distinguished fellow at Stanford University's Hoover Institution*

When we talk about energy, what we are really talking about is our prerogative as an exceptional society. The fuels we use and our machines that harness them reflect our ambitions for progress and our values around how we get there. New technologies, the result of research and development that may otherwise seem esoteric, enable that pursuit. And the more choices our laboratories put on the table, the less constrained we are in using them to reach the things we really care about—health, family, business, culture, faith, and delight. This is what game changers are ultimately about.

Today, the average American commands nearly fifty times the energy that a citizen of the Roman Empire did. We command five times that our forefathers did just a hundred years ago. Over the same time, the amount of economic activity we create per unit of energy available to us has tripled. A Model T produced 20 horsepower; today's Ford family sedans produce 200—at the same adjusted sales price, and with twice the mileage. A Boeing 777 produces over 200,000 horsepower. Our prosperity surely reflects this. But, we still spend nearly $1 billion per day on oil imports, lose $100 billion annually to power grid outages, and grapple with environmental and health impacts of energy use for which damages may even exceed the price we pay out of pocket.

Many US soldiers have been killed in Iraq and Afghanistan protecting fuel convoys. In a perfect world, we would not need to talk about energy at all. Energy would become transparent, like air. We have a great start, but we are not yet there, so we need to keep pushing, and keep energy on the move.

ABOUT THE CONTRIBUTORS

Robert C. Armstrong is director of MITEI. He was cochair of the Energy Research Council that laid the groundwork for MITEI and served as deputy director for the initiative's initial six years, during which time it funded more than eight hundred research projects. Armstrong is the Chevron Professor of Chemical Engineering and has been a member of the MIT faculty since 1973. He headed the Department of Chemical Engineering from 1996 to 2007. His research interests include polymer fluid mechanics, rheology of complex materials, and energy. In 2008, Armstrong was elected into the National Academy of Engineering for conducting outstanding research on non-Newtonian fluid mechanics, coauthoring landmark textbooks, and providing leadership in chemical engineering education.

Jeremy Carl is a research fellow at the Hoover Institution and director of research for the Shultz-Stephenson Task Force on Energy Policy. His work focuses on energy and environmental policy, with an emphasis on energy security and global fossil fuel markets. Before coming to Hoover, Carl was a research fellow at the Program on Energy and Sustainable Development at Stanford and a visiting fellow in resource and development economics at the Energy and Resources Institute in New Delhi, India. His writing and expertise have been featured in the *New York Times, Wall Street Journal, Newsweek,* and many other publications. He holds degrees in history and public policy from Yale and Harvard Universities.

Louis Carranza is associate director of MITEI, which he joined in 2013. He was previously vice president for strategic development at IHS, where he worked for seventeen years. At IHS, he served as executive director and cochair of CERAWeek—one of the top five corporate leader conferences in the world—and conceived of and implemented the CERAWeek partnership program. He was also codirector of the IHS scenario planning initiative. Before joining IHS, Carranza managed Cambridge Energy Research Associates' global power practice.

David Fedor is a research analyst on the Hoover Institution's Shultz-Stephenson Task Force on Energy Policy. He has worked in energy and the environment across China, Japan, and the United States. Formerly at the Asia Pacific Energy Research Center and Stanford's Collaboratory for Research on Global Projects, Fedor has also consulted for WWF China, the Asian Development Bank, and the Korea Energy Economics Institute. He holds degrees in earth systems from Stanford University.

Rebecca Marshall-Howarth is MITEI's editorial services and publications director. After two decades in the computer industry, focusing on process control and the oil, gas, and power markets, she joined MITEI in 2009. She provides editorial assistance to both staff and faculty authors of MITEI publications. She also oversees the design and production of MITEI publications, in addition to serving as custodian of the MITEI brand. She was the key contact for copyediting, design, and production of this book.

Francis O'Sullivan is MITEI's director of research and analysis and a lecturer at the MIT Sloan School of Management. He has been at MITEI since its early days, most recently serving as the executive director of the Energy Sustainability Challenge. His current research focuses on solar, unconventional oil and gas resources, and the energy-water nexus. O'Sullivan is a member of the National Academies' Roundtable on Science and Technology for Sustainability. Previously, he was a consultant at McKinsey & Company in the areas of economic, investment, and risk analysis; strategic planning; and operations in the private equity, oil and gas, electric utility, and renewable energy sectors.

George Pratt Shultz is the Thomas W. and Susan B. Ford Distinguished Fellow at the Hoover Institution. He is one of two individuals who have held four different cabinet posts; has taught at three of this country's great universities; and for eight years was president of a major engineering and construction company. Shultz was sworn in on July 16, 1982, as the sixtieth US secretary of state and served until January 20, 1989. He is advisory council chair of the Precourt Institute for Energy at Stanford University, chair of the MIT Energy Initiative External Advisory Board, and chair of the Hoover Institution's Shultz-Stephenson Task Force on Energy Policy.

David Slayton is a Hoover Institution research fellow. He originally came to Hoover as a US Navy national security affairs fellow after his command tour with the Electronic Attack Squadron 134. During his career, Slayton deployed twelve times and participated in a broad range of combat operations on the ground, at sea, and in the air, conducting numerous combat missions into Afghanistan and Iraq. A career naval flight officer, he has flown more than three hundred combat missions. His research focuses on the electromagnetic spectrum, maritime strategy, future energy sources, transmission, and infrastructure. Slayton holds degrees from the University of California at Los Angeles, the University of San Diego, and the Naval War College.

Energy Policy

The Hoover Institution's Shultz-Stephenson Task Force on Energy Policy addresses energy policy in the United States and its effects on our domestic and international political priorities, particularly our national security.

As a result of volatile and rising energy prices and increasing global concern about climate change, two related and compelling issues—threats to national security and adverse effects of energy usage on global climate—have emerged as key adjuncts to America's energy policy; the task force will explore these subjects in detail. The task force's goals are to gather comprehensive information on current scientific and technological developments, survey the contingent policy actions, and offer a range of prescriptive policies to address our varied energy challenges. The task force focuses on public policy at all levels, from individual to global. It then recommends policy initiatives, large and small, that can be undertaken to the advantage of both private enterprises and governments acting individually and in concert.

Contact for the Shultz-Stephenson
Task Force on Energy Policy:
Jeremy Carl, *Research Fellow*
(650) 723-2136
carljc@stanford.edu

MIT Energy Initiative

MITEI works to help transform global energy systems. It is a research, education, and outreach program that, in its depth and breadth, is without peer at US academic institutions. An institute-wide initiative, MITEI pairs MIT's world-class research teams with key players across the innovation spectrum to help improve today's energy systems and shape tomorrow's global energy marketplace. It is also a resource for policy makers and the public, providing unbiased analysis and serving as an honest broker for industry and government.

MITEI has more than sixty-eight industry and public partners and has funded more than 128 novel or early-stage energy research projects submitted by faculty from across MIT.

MITEI's educational offerings combine single-discipline depth with multidiscipline breadth, transforming the MIT campus into an energy learning laboratory. The energy studies minor, established in 2009, is among the largest minors at MIT.

Contact for the MIT Energy Initiative:
Frank O'Sullivan
Director of Research and Analysis
(617) 715-5433
frankie@mit.edu

INDEX

silicon
  high-efficiency monocrystalline
    rear-junction silicon cells,
    20–21
  nanocrystalline-silicon shells,
    23–24
  nanoparticle batteries, 54–55
  in PV, 19–21
SLAC National Accelerator Laboratory,
    of DOE, 44–45, 91
slick water fracturing, 1, 3, 5, 6, 77–78
smart oil fields, 6–7
Smart Voyage Planning (SVP), 97
software, 85–99
solar power. *See also* photovoltaics
Soldier Wearable Integrated Power
    System (SWIPES), 101
solid oxide fuel cells, 38
Soljacic, Marin, 60
Somerville, Chris, 92
Sony, 50
Squad Power Manager, 97, 101
Stanford
  ambient seismic testing by, 11–12
  catalysts by, 91
  computational chemistry by, 91
  copper hexacyanoferrate battery,
    43–44
  D. School at, 66
  graphene in fuel cells by, 39–40
  Hoover Institution at, xiv
  hydrogel lithium ion batteries
    by, 54–55
  induction charge system by, 60–61
  lithium polysulfide membrane-free
    flow battery, 44–45
  nanocrystalline-silicon shells
    by, 23–24
  polariton laser by, 72–73
  silicon nanoparticle batteries
    by, 54–55
  synthetic magnetism photon
    control by, 73–74
  thermoelectrics and thermionics
    at, 79
stern flaps, for Navy, 95–96

stress corrosion cracking, in nuclear
    power plants, 84
SunPower Corporation, 20–21
Sunshot team, of DOE, 25
superlattice materials, for fuel
    cells, 38–39
surfaces, at nanoscale, 81–84
SVP. *See* Smart Voyage Planning
Swanson, Richard, 20–21
Swanson's Law, 21
Sweatman, Peter, 61
SWIPES. *See* Soldier Wearable
    Integrated Power System
synthetic magnetism photon
    control, 73–74

tactical vehicle fuel efficiency, 101–02
tandem solar cells, 26
Tesla Model S, 50, 52
thermionics, 78–81
thermoelectrics, 78–81
thin-film LED, 71
thin-film organic polymer flexible
    solar cells, 27–28
Thompson, Levi, 58
3D seismic imaging, 4–5
TILES, 7
Tinker, Scott, 9

ultracapacitors, 49
ultradeepwater offshore drilling, 2
Union Pacific Railroad, 5
Union Pacific Resources, 77
University of California Berkeley
  catalysts by, 92
  computational chemistry by, 92
  Energy Biosciences Institute at, 92
  thermoelectrics at, 80–81
  zinc batteries by, 57–58
University of Michigan
  driverless vehicle test bed by, 61–62
  fluorescent replacements by, 65
  organic, tandem PV by, 28–29
  thermoelectrics at, 79–80
  vanadium nitride supercapacitors
    by, 58–59